BRAIN WARS

ALSO BY MARIO BEAUREGARD

The Spiritual Brain:
A Neuroscientist's Case for the
Existence of the Soul

BRAIN WARS

The Scientific Battle Over
the Existence of the Mind and the
Proof That Will Change
the Way We Live Our Lives

Mario Beauregard

HarperCollins*Publishers*Ltd

Published by HarperCollins Publishers Ltd

First Canadian edition

HarperCollins books may be purchased for educational, business,
or sales promotional use through our Special Markets Department.

HarperCollins Publishers Ltd
2 Bloor Street East, 20th Floor
Toronto, Ontario, Canada
M4W 1A8

www.harpercollins.ca

Library and Archives Canada Cataloguing in Publication
Beauregard, Mario, 1962–
Brain wars : the scientific battle over the existence of the mind
and the proof that will change the way we live our lives / Mario
Beauregard.

ISBN 978-1-44340-706-9

1. Consciousness. 2. Brain. I. Title.
QP411.B42 2012   612.8'2   C2012-900716-1

Printed and bound in the United States
RRD 9 8 7 6 5 4 3 2 1

CONTENTS

A "Computer Made of Meat"?

Materialism and the Mind-Brain Problem

If . . . Descartes . . . had kept a poodle, the history of
philosophy would have been different. The poodle would
have taught Descartes that contrary to his doctrine,
animals are not machines, and hence the human body is
not a machine, forever separated from the mind . . .

—ARTHUR KOESTLER[1]

Very often, perhaps more often than you think, ordinary people
have unexplained experiences that defy the boundaries and explanations of traditional science:

A group of patients suffering from painful osteoarthritis of
the knee—some leaning on canes to walk, and all diagnosed to
be in need of knee surgery—participate in a double-blind experiment in which some are given real knee surgery, and others
are given placebo, or "fake," surgery. The results are startling: *all*
of the patients report much less pain, and those who had gotten

the placebo surgery—that is, no surgery—are even able to walk and play basketball.

One evening after dinner, while watching the sunset from his living room, a successful physician and award-winning research psychiatrist enters into a mystical state of illumination in which time seems to disappear. He returns to ordinary consciousness a few minutes later, the course of his life changed forever.

Maria, a migrant worker, has a severe heart attack and is admitted to a hospital coronary unit, where medical personnel rush to restart her heart and save her life. Later, she tells a social worker that during her cardiac arrest she looked down on her body from the ceiling and watched the doctors work. And while she was there, she looked out the window and spotted a tennis shoe on the ledge of another building. Curious, the social worker searches for the tennis shoe—and finds it, right where Maria said it would be—in a spot that Maria could not possibly have seen from her hospital bed.

What happened?

Were the knee patients who received the placebo surgery healed by the power of their belief? Or is arthroscopic surgery simply not all it is cracked up to be?

Did the psychiatrist really experience what Aldous Huxley called "mind at large"? Did his consciousness fuse with the source of the universe? Or was he simply hallucinating or having a waking dream?

Did Maria leave her body and float to the ceiling? And how could she see a tennis shoe that was hidden from her view?

Is any of this really possible?

The questions we are asking in this book are age-old: Are we nothing but sophisticated animals? Where does our sense of self originate? *Is there a difference between our brains and our minds or consciousness?* What happens to us after our body perishes? Does our consciousness completely dissolve into nothingness? Can the mind exist without the body?

Over the past several centuries in the West, most scientists have functioned within a strict materialist framework that holds to one essential assumption: matter is all that exists. This materialist viewpoint has become the lens through which most of us interpret the world, interact with it, and judge what is true. Mainstream neuroscientists—scientists like myself, who study the brain and how it works—operate from the view that electrical impulses in the brain account for all of our thought patterns and mental experiences. In the harsh judgment of the famed molecular biologist Francis Crick, the co-discoverer of the molecular structure of DNA, "'You,' your joys and your sorrows, your memories and your ambitions, your sense of personal identity and free will, are in fact no more than the behavior of a vast assembly of nerve cells and their associated molecules."[2]

Yet . . . is matter *really* all that exists? Most of us, regardless of our religious beliefs, feel intuitively that our consciousness—our selves, what makes us unique individuals—resides apart from our bodies. Human cultures globally and throughout history have conceived of a vital principle, the seat of mind and self, that survives the death of the body as an individual spirit or as a part of a Universal Spirit. For the past several centuries scientists have had to put their innate beliefs to one side in order to do the fine, objective work that has enabled so many remarkable discoveries. We have gained much through this method, but along the way science seems to have lost sight of a wider view: the open door to the possibility of the "impossible."

Within the view of materialism, everything is composed of collections of material particles. All that we experience—including our thoughts, feelings, beliefs, intentions, sense of self, and spiritual insights—results from electrochemical impulses in our brains. In this world, people who report transcendent experiences—from out-of-body experiences like Maria's, or clairvoyant visions of things we can't possibly know, or ecstatic experiences in which we seem to merge with the universe—must by definition be hallucinating or having some sort of momentary brain malfunction.

For me, for a growing number of scientists, and perhaps for you too, this sort of thinking does not resonate with what we feel and experience every day, let alone in moments that transcend the ordinary. In fact, materialism's truncated understanding of what it means to be human often prevents us from seeing what is staring us in the face. Physician and author Dr. Larry Dossey puts his finger on the core issue: "There's [a dreary] little secret that veteran scientists never let kids in on—that if they enter science, they have to check their minds at the door. The reason is that mind, as most people think about it, does not exist in conventional science because the expressions of consciousness, such as choice, will, emotions, and even logic are said to be brain in disguise."[3]

We have been checking our minds at the door for a long time. Yet many of us, whether we describe ourselves as religious or even spiritual, look around at the intricate workings of nature and our own bodies and feel innately that "there must be more." Yes, we understand that our brains direct the amazing systems that keep our bodies running; at the same time, we stubbornly continue to believe that "we"—the intangible stuff that we identify as ourselves—are more than our brains.

IIII

The brain can be weighed, measured, scanned, dissected, and studied. The mind that we conceive to be generated by the brain, however, remains a mystery. It has no mass, no volume, and no shape, and it cannot be measured in space and time. Yet it is as real as neurons, neurotransmitters, and synaptic junctions. It is also very powerful.

A wealth of scientific studies indicate that our thoughts, beliefs, and emotions influence what is happening in our brains and bodies and play a key role in our health and well-being. For better or for worse, we create our lives, our health, and our worlds with our minds. You will read about many of these studies in the chapters of this book. Research shows the real and often untapped power of the mind.

Our belief in the efficacy of a bogus treatment—a treatment proved by science not to work—can stimulate our self-healing capacity even in diseases as severe as cancer and Parkinson's disease. Emotions can even affect whether certain genes are turned "on" or "off"—altering, for example, our bodies' response to stress.

We can deliberately change processes in our bodies that are normally not under voluntary control. We can improve our cognitive performance by learning to control the electrical activity of our brains using neurofeedback. We can train our minds—through meditative practices—to enhance the activity of brain areas implicated in emotional well-being, compassion, and attention. And mental training can even change the structure of our brains.

Hypnosis can enable people to see color where there is only gray. Brain wave receptors can enable people to move objects using only their thoughts. The impossible is indeed possible.

Along with an increasing number of scientists, I believe vehemently that *the materialist framework is not science*. A growing body of solid scientific research shows this view to be not only limited, but wrong. In fact, few mainstream scientists operating within the paradigm of scientific materialism realize that it is based on a number of philosophical assumptions—that is, *beliefs without proof*. At least three of these basic assumptions can be traced back to classical physics:

> *Physicalism* is the idea that only matter and energy exist in the universe. This means that life, mind, and consciousness are simply accidental by-products of complex arrangements of matter (and energy).

> *Reductionism* is the notion that complex things can be understood by reducing them to the interactions of their parts, or to simpler or more fundamental things. Key clues to a reductionist viewpoint include terms like *just, merely,* and *nothing but.*

> *Objectivism* is the thesis that scientists should investigate empirical facts in an objective manner: using the physical senses and their technological extensions, including such instruments as microscopes and telescopes.

The story of how science came to accept these ideas without question is invaluable in helping to understand not only why they have been useful for so many years, but why the time has come for science to begin thinking outside the box that originated in ancient Greece.

IIII

The Greek philosopher Heraclitus was the first to suggest that human beings consist of body and soul, and that the soul is responsible for thinking and emotions. A bit later, Plato took this idea further and proposed a *dualistic* view: the body is the temporary receptacle of this immortal and invisible soul, which is bound to the brain.[4] (Remember this view—we will come back to it again in our story.)

Democritus, a bit younger than Plato, was skeptical of such supernatural explanations. He proposed a theory that may sound quite familiar: all that exists is composed of matter and void; the world is made up of basic material particles (which he called "atoms," meaning "indivisibles"); and humans are material beings that cease living when the body stops functioning.

Democtritus's ideas were something of a sea change in the mind/brain debate. His contemporary, Hippocrates—the "father of Western medicine"—took this theory a step further. Brain injury can impair mental functioning, he argued, so the brain must be the seat of consciousness, intellect, and emotions. A few centuries later, in the second century, the Roman physician and philosopher Galen postulated that mental capacities such as perception and reasoning are dependent on the brain.[5]

During the Middle Ages in Europe, the church had become a powerful force, permeating everything. "Science" was alchemy, a murky brew of religion and superstition. Free observation and exploration was discouraged, and little was known about the causes and cures of disease. During the Renaissance in the fifteenth and sixteenth centuries, scientific interest apart from religion blossomed once again. And by the seventeenth century, the Scientific Revolution was in full swing.

The French philosopher and mathematician René Descartes revived Plato's conception of a mind/body dualism. Most famous for his pronouncement "I think, therefore I am," Descartes saw human beings as an amalgamation of material bodies and immaterial minds, both of which had their source in God. Bodies, in his view, are machines that function in accordance with the laws of physics, but minds are exempt from those laws because they are not physical. Descartes also theorized that the mind acts willfully upon the body. Most of the prominent philosophers and scientists who followed Descartes rejected his theory because it failed to satisfactorily account for the interaction between mind and brain. How, they asked, can an immaterial, mental substance act upon the material brain?

The Scientific Revolution brought groundbreaking discoveries that swept away superstition: the laws of gravity and the laws of planetary motion showed us how the world worked; calculus showed us that the world could be measured and calculated; the microscope showed us invisible worlds never imagined. These tools—and the invention of scientific method itself—laid the foundations of modern science and gave us the modern world.

With the Enlightenment in the eighteenth century, religious revelations were increasingly challenged[6]; rationality and science were advocated as the primary sources for authority and legitimacy. Advances in technology allowed scientists to peer into blood and water and other elemental substances to discover teeming life forms. As they began to identify various agents of diseases, such as bacteria, fungus, and viruses, physicians understandably began to believe that the treatment of infectious diseases required only the elimination of such agents.

The Industrial Revolution, with the stunning power of machines to reshape the world and the revolutionary possibility of "progress," gave science a new concept with which to reframe its thought. The body was no longer seen as having been created on a divine model—it was a *biological machine*. Food in, energy out. In this world, medicine's role was no longer to intercede with God for mercy; it was purely mechanical, to repair bodily malfunctions.[7]

With new insights into how the smallest elements of life worked, old concepts of causation began to dissolve. Astronomical and other observational insights led to new philosophies describing a universe that was nothing more than a cold, impersonal, deterministic machine composed of mindless bits of matter, devoid of consciousness and intelligence, and bereft of purpose. Darwin's theory led some to wonder whether humanity was in fact something of an evolutionary accident. The German philosopher Friedrich Nietzsche proclaimed God dead, and other modern philosophers professed that supernatural beliefs about spiritual beings were nothing more than chimeras of religious fantasy.[8]

During the second half of the nineteenth century substantial progress was made on the study of the impact of brain damage on mental functions, such as language, memory, reasoning, and perception. This dealt a blow to the idea of dualism. The prevailing view among scientists was that the causes of mental activities and behaviors were found solely within the brain—and it seemed to have been proven in the laboratory: human consciousness no longer belonged to an inexplicable realm. In 1874, the English biologist Thomas Huxley published an article suggesting that human beings are conscious biological machines.[9]

By the beginning of the twentieth century, this materialist view dominated science. Nonetheless, some philosophers and scientists resisted the materialist trend. In 1891, Oxford philosopher Ferdinand Schiller proposed that matter is not what produces consciousness but what limits it.[10] In 1898, William James[11]—the father of American psychology—pointed out the fact that scientists can only measure *correlations:* when brain states change in a certain fashion, mental states change too. The fact that mental functions are disturbed when the brain is damaged does not prove that the brain generates mind and consciousness.

Using an elegant analogy we will return to again in this book, James explained what he meant: when white light passes through a prism, he said, the prism allows it to be broken up into all the colors of the spectrum. *The prism is not itself the source of the light,* but it permits us to see the light differently. In the same way, the brain may permit, transmit, and express mental events and conscious experiences that have their source elsewhere. It does not produce them. James felt that this hypothesis could also account for the effects of drugs and brain damage.[12]

To date, a number of scientists and thinkers have used an updated form of James's analogy to illustrate the mind-brain relationship: equating "mind" with "brain" is as illogical as listening to music on a radio, demolishing the radio's receiver, and thereby concluding that the radio was creating the music.

||||

During the twentieth century, mind-brain dualism made a quiet return to scientific circles. This was thanks primarily to the work of researchers Wilder Penfield and John Eccles.

In 1934, Canadian neurosurgeon Wilder Penfield became the first director of the renowned Montreal Neurological Institute. Ironically, he began his career as a scientist with the intention of proving the materialist thesis that mind is simply the product of brain activity. To do so, he developed a surgical procedure in which he treated patients with severe epilepsy by removing cerebral tissue that caused epileptic seizures. Before destroying the epileptic tissue, Penfield stimulated various regions of the brain with electrodes to map cerebral functions. During this procedure patients lay on the operating table fully conscious and alert—they could not feel any pain because brain tissue has no sensitivity.

Penfield performed brain mapping in over a thousand patients during the course of a few decades. He discovered that most of the time, electrical stimulation of the brain elicits rudimentary sensations or motor responses; very rarely, it elicits hallucinated images or scenes. Strikingly, Penfield's patients were always able to discriminate between mental events they had willed and those that were evoked by the electrical stimulation. For instance, if an electrode applied over the motor area of a patient's brain induced a simple hand movement, the patient would tell Penfield, "I didn't do it. You made me do it."[13] At the end of his scientific career, Penfield concluded that higher mental functions—such as consciousness, reasoning, imagination, and will—are not produced by the brain: mind is a nonphysical phenomenon interacting with the brain.[14]

John Eccles, an Australian neurophysiologist, was awarded the 1963 Nobel Prize in Physiology or Medicine for his work on the synapse, the tiny junction across which a nerve impulse passes from one neuron to another neuron. Eccles felt that materialist theories fail to account for the wonder of being human: "I maintain that the human mystery is incredibly demeaned by scientific reductionism," he said.[15] Eccles's dualistic interactionist theory claims that humans have a nonmaterial mind—belonging to the mental world—that acts upon, and is influenced by, the material brain—which belongs to the physical world. Contrary to Descartes, Eccles argues that the nonmaterial mind is not a substance. Furthermore, he postulates psychic-to-physical laws that regulate the effect of the mind on the brain.[16]

||||

Penfield and Eccles, despite acclaim for their work, were out of the mainstream. Scientists during the twentieth century formulated a number of materialist positions about the mind-brain problem, notably mind-brain identity theory, eliminativism, and emergent mentalism.

Mind-brain identity theory asserts that mental events are created by, and are identical to, brain events.[17] Although we experience our own consciousness and mental events subjectively ("I feel great"), we can measure events in our brain from the outside, in an objective manner ("My brain released endorphins after my run"). This theory implies that our sense of identity, our beliefs and values, our feelings, our spiritual experiences, and even our free will are nothing more than electrical impulses and chemical reactions in our brain.

Eliminativism, a more radical position, resolves the mind-body problem by denying the existence of mental functions and events, and claiming that there are only physical states of the brain. According to this view the mental world is an illusion, and we only imagine ourselves to have thoughts, memories, feelings, and goals. Consciousness, mind, free will, and self are pre-scientific concepts belonging to unsophisticated ideas (dubbed "folk psychology") of how the brain works. Eliminativism holds that these concepts will someday be eliminated, thanks to the progress of science.[18]

Emergent mentalism, proposed by Nobel laureate and neuro-scientist Roger Sperry, contends that mental events are not possible independently of the brain. Yet Sperry speculated that thoughts, beliefs, intentions, feelings, desires, moral values, and other mental events are higher-level properties that emerge from (but cannot be reduced to) lower-level physical processes in the brain. He also postulated that such mental events can causally influence the brain processes that create them.[19] Various formulations of emergent mentalism have appeared following Sperry.[20]

None of these theories explain what the Australian philosopher David Chalmers has referred to as the "hard problem" of consciousness:[21] why and how subjective inner experiences, such as love and spiritual epiphanies, arise from physical processes in the brain. Given that the evidence from neuroscience does not demonstrate that all mental events, without exception, are caused solely by neural processes, a growing number of people, including Chalmers himself, now question whether this problem will ever be solved by materialist theories of the mind.

Materialism remains the dominant scientific view. Artificial intelligence pioneer Marvin Minsky memorably encapsulated the materialist position: "The brain," he said, slightly updating the machine paradigm, "is just a computer made of meat." Minsky, however, had not reckoned with the infinite worlds of possibility being opened up by a revolutionary new branch of physics called quantum mechanics (QM). QM has effectively smashed the scientific materialist worldview and, as we will see later in this book, is opening up infinite new worlds of possibility that are blowing the materialist box to smithereens.

▌▌▌▌

Materialist theories, despite their stubborn persistence in the scientific community, cannot solve the mind-brain problem.[22] We need a new model through which to view the power of mind and its central role in the universe. This fresh point of view would be free of materialist dogmas that have outlived their time and blocked science from exploring avenues that have been there all along.

Like Frederic Myers (a pioneer researcher in the psychology of the unconscious, or "depth psychology," who greatly influenced William James and the Swiss psychiatrist Carl Gustav Jung), I believe that great progress can be made if we follow a pillar of the scientific method and approach the mind-brain problem empirically:[23] giving priority to the knowledge derived from observations and evidence.[24] In the chapters that follow you will find a large body of evidence related to the impressive capacities of the mind and its fundamental, irreducible, nature.

As you will see, multiple lines of hard evidence show that mental events do exist and can significantly influence the functioning of our brains and bodies. They also show that our minds can affect events occurring outside the confines of our bodies, and that we can access consciously transcendent realms—even when the brain is apparently not functioning. Most important, these various lines of evidence indicate that materialist theories of the mind are erroneous: we are not merely complex biological machines, computers made of meat. Reality is a vastly complex territory that we are only beginning to explore. It encompasses, as you will see in the chapters that follow, much more than the physical world.

Times are changing quickly, particularly in science. The most exciting frontiers of twenty-first century science—quantum mechanics, cloud computing, virtual reality—show us very different models of what is real and what is possible than materialist science permits. They also give us tools with which we can explore the nature of the relationship between our minds—our consciousness, our self-identity—and our brains. Great advances in science are made by following the evidence, wherever it may lead. If our objective is truly to reach an adequate scientific understanding of the human mind, then we must be willing to take into account all the empirical evidence related to this issue. This is what *Brain Wars* sets out to do, and what I hope you will do as you read this book.

I

The Power of Belief to Cure or Kill

The Placebo/Nocebo Effect

Placebos are the ghosts that haunt our house of biomedical
objectivity, the creatures that rise up from the dark and
expose the paradoxes and fissures in our own self-created
definitions of the real and active factors in treatment.

—MEDICAL HISTORIAN ANNE HARRINGTON[1]

In the late 1950s, a patient whom psychologist Bruno Klopfer
called "Mr. Wright" was on his deathbed in a hospital in Long
Beach, California, dying from an advanced cancer of the lymph
nodes. That he was dying was indisputable: large tumor masses,
the size of oranges, had proliferated in his neck, groin, chest, and
abdomen. His spleen and liver were huge, and between thirty
and sixty ounces of fluid had to be drawn from his chest every
other day just to enable him to breathe. His doctors expected
him to live not more than two weeks.[2]

But Mr. Wright was not ready to die. When he learned that his treating physician, Dr. Philip West, was involved in the testing of a new cancer drug, he was filled with hope. This new drug, called Krebiozen, had initially appeared to be very effective; he begged Dr. West to give him the revolutionary drug. Touched by his patient's desperation, and knowing that his case was hopeless, West agreed.

Dr. West gave Mr. Wright an injection of Krebiozen on a Friday afternoon. The following Monday, the doctor was amazed to find his "dying" patient ambling around the unit and joking with the nurses. And he not only seemed better, he *was* better. Over the weekend, the man's tumors had melted like snowballs on a hot stove. After ten more days of treatment with the experimental drug, practically all signs of his disease had vanished. In what seemed like a triumph for Krebiozen, Mr. Wright was discharged from the hospital and returned to his normal lifestyle, seemingly cured.

Two months later, however, Mr. Wright opened the newspaper and read an article that sent his spirits plunging. According to preliminary reports, researchers had found that Krebiozen was *not* effective for the treatment of cancer. Despite his previous recovery, Mr. Wright suffered an immediate relapse and his tumors came back.

Deeply affected by this turn of events, Dr. West resorted to a desperate trick—lying. His voice filled with conviction, Dr. West told his patient that the newspapers were wrong and that Krebiozen was a powerful drug against cancer. He explained to Mr. Wright that his relapse was due to the fact that the dose he had been administered came from a batch that had deteriorated while in the pharmacy, but that he had just received a new "super-refined, double strength" batch of the drug that was capable of duplicating the initial healing effects. Convinced by this story, Mr. Wright eagerly agreed to begin a second treatment program.

Full of expectation and faith, Mr. Wright watched as he was injected with fresh distilled water that he believed was the miracle drug Krebiozen. The effects were as astonishing as they had been with the first injections of the drug. Again the tumor masses melted, chest fluid disappeared, and within a few days he walked out of the hospital, fully recovered and symptom free.

He remained in this happy state until a few months later, when he read a newspaper report announcing the final verdict of the American Medical Association. This report confirmed the earlier findings: Krebiozen was totally ineffective and useless against cancer. Within a few days of reading this article, Mr. Wright was readmitted to the hospital in a very distraught state. Two days later, he died.

||||

A few years ago, *New Scientist* magazine published a paper titled *13 Things That Don't Make Sense*,[3] listing a series of scientific anomalies, phenomena that just don't make sense according to the accepted wisdom. The placebo effect—our ability to heal ourselves or alleviate pain by simply believing in the treatment, whether or not it is "real"—was ranked number one on the list. As with Mr. Wright, it seems that the more firmly we believe in a particular treatment—whether it's a sugar pill, an injection of salt water, a shaman's curse, or an FDA-approved medication— the more likely it is that the treatment will work.

Some materialistically inclined scientists see the placebo

effect as an anomaly and attempt to dispute its existence. For instance, Danish researchers Asbjørn Hróbjartsson and Peter C. Götzsche found certain flaws in older studies involving placebo treatments. Based on this information, they argued that there is little evidence that such treatments can produce real, lasting effects apart from normal disease processes.[4]

But is the placebo effect really an anomaly? A body of research, including my own, indicates strongly that mental states such as beliefs and expectations can be very powerful.[5] The placebo effect and its evil twin, the nocebo effect—the unpleasant and even life-threatening symptoms that can follow the administration of a fake treatment—have been known for centuries. These effects are now accepted as undeniable by the vast majority of physicians.

The power of mind over body is not just myth, wishful thinking, or witchcraft—although these, it seems, may be genuinely useful triggers for healing. Placebo effects—even those as startling as the disappearance of tumors overnight—appear abnormal only if you assume that "mind" is an illusion produced by the workings of the brain. Yet science very definitely tells us otherwise.

Placebo is a Latin expression that means "I will please." It appeared in the Bible following Saint Jerome's mistranslation of the first word of the ninth line of Psalm 116. Instead of translating the Hebrew "I will walk before the Lord," he wrote, *Placebo Domino in regione vivorum* ("I will please the Lord in the land of the living").[6] In fact, the first placebos were people, not pills. In the Middle Ages, when professional paid mourners waited for Vespers for the Dead to begin, they often chanted the ninth line of Psalm 116 and so were called "placebos" to describe their fake lamentations.

Placebo controls—the administration of fake procedures to separate the effects of imagination from reality—began in Europe in the sixteenth century with progressive Catholic efforts to discredit exorcisms. Individuals supposedly possessed by diabolical forces were given false holy objects. If the person reacted with violent contortions, officials concluded that his possession was in his imagination.[7]

At the end of the eighteenth century, the word placebo was being used in a medical context much as we do today: to describe harmless treatments aimed at supporting the healing process. Until the 1850s, doctors did not have efficient methods to combat disease, and placebo was the main treatment they had to offer to their patients. It was enough, these doctors found, that patients believed in the cure—even if the doctors did not.

Like so many leaps in understanding, the place of the placebo effect in science was rooted in a battlefield discovery. Toward the end of World War II, a nurse was assisting an American anesthetist named Henry Beecher in a military field hospital located in southern Italy. One day the morphine supply ran low because of heavy casualties. This could not have happened at a worse time—Beecher had to treat a badly wounded soldier in need of an urgent operation. The nurse thought quickly. Though her syringe contained only salt water, she told the soldier that he would receive an injection of a powerful painkiller. To Beecher's amazement, the bogus injection rapidly relieved the soldier's condition, and he did not appear to feel significant pain during the ensuing surgery.

After the war Beecher returned to his post at Harvard, but he did not forget what he had seen. In 1955 he published an article titled "The Powerful Placebo" in the *Journal of the American Medical Association* (JAMA).[8] In it, he described how the placebo effect had contaminated the results of several pharmaceutical trials by causing improvement that was mistakenly attributed to the drugs being tested. He further claimed that about 35 percent of patients responded positively to placebo treatment, and argued that trial volunteers who received real medication were subject to placebo effects. Because the act of taking a pill was itself somehow therapeutic, Beecher proposed that only by subtracting the amelioration in a placebo control group from the total therapeutic effect of the drug being tested could the real value of this drug be calculated.

In 1962, following the publication of Beecher's influential article, the U.S. Congress required drug trials to include placebo control groups. Volunteers would be assigned randomly to receive either the medicine being tested or a sugar pill, and neither researcher nor patient would know the difference until the trial was over. Beecher's double-blind, placebo-controlled, randomized clinical trial was consecrated as the gold standard of the emerging pharmaceutical industry. Today, this standard is firmly entrenched: a new medication must beat placebo in at least two authenticated trials to win U.S. Food and Drug Administration (FDA) approval.

But Beecher had encouraged an inflated notion of the placebo effect by failing to distinguish the placebo response from other confounding factors, such as random fluctuation of symptoms and spontaneous remission.[9] Moreover, most of the early studies cited in Beecher's article did not include a control group that received no treatment. Without such a group, it is not possible to exclude the possibility that the improvement shown by the members of the placebo group might have occurred anyway, even if they had not received any placebo.[10]

Since the 1990s, a number of studies have investigated the placebo effect by using objective physiological measurements and no-treatment groups. Overall, the results of these studies indicate that the belief that one has been given some potent treatment and the belief in the therapist, whether a Western-trained doctor, an acupuncturist, or a traditional healer, appear to mobilize powerful innate self-healing mechanisms.[11]

Belief, of course, is culturally specific. If you are an American believer in the power of Western medicine, a hypodermic needle, a striped capsule, or a sophisticated machine with several buttons may be the symbol of a powerful treatment. A member of a remote indigenous tribe might find the ingestion of a specific plant picked at the full moon much more credible.[12] And regardless of the treatment, you may experience symptom relief. Because *anything* that increases the credibility of a particular type of treatment will enhance the capacity of that treatment to evoke a placebo response.

And by anything, I really do mean anything. Mental processes and social contexts that facilitate faith and hope, positive expectations and feelings, motivation, and anticipation of improvement also appear to affect our brains, bodies, and behavior. To a certain extent, a caring attitude on the part of the therapist toward the patient may also contribute to the induction of a placebo response.

In clinical trials, geographic location appears to influence whether a drug can beat a placebo. For example, when the antidepressant drug Prozac was tested, it performed better in the United States than it did in Western Europe and South Africa. In line with this, Daniel Moerman, a professor of medical anthropology at the University of Michigan–Dearborn, found that Germans are low placebo reactors for hypertension—a medical condition undertreated in Germany—but high in trials of drugs for ulcer.[13] And even the attributes of pills—size, color, shape, and even price—influence their effects on the body.[14]

IIIII

Most children learn early on that if they scrape their knee, Mommy can "kiss it and make it better." In the late 1980s, a compelling study about the reduction of the sense of pain triggered by a belief in a fake treatment was conducted with patients who had undergone tooth extraction.[15]

Ultrasound is often used after tooth extraction to diminish postoperative pain and inflammation, and to accelerate healing. In this experiment, four to six hours after surgery, groups of patients received no therapy, ultrasound therapy, or "mock" ultrasound therapy. The study was double-blind as it was impossible for the researchers or the patients to discern whether or not the machine was emitting the sound waves, which are inaudible to the human ear. Researchers measured facial swelling twenty-four hours after surgery; and patients were asked to indicate their level of pain on a scale ranging from "no pain" to "unbearable pain." The results were quite interesting.

The groups treated with the ultrasound machine (both on and off) reported a significant decrease in pain, compared with the untreated control group. Remarkably, the patients who had been treated with the machine while it was turned off reported the same level of pain reduction as those who had received the proper ultrasound treatment. Furthermore, swelling was significantly reduced in those who received the fake ultrasound compared to the no-treatment control group. Thanks to such studies, scientists now have a fairly good understanding of how placebos alleviate pain. The first double-blind study that specifically addressed this issue was conducted by Jon Levine and his colleagues in 1978.[16] These researchers administered a placebo as a painkiller to forty patients following dental surgery. The patients reported a reduction in pain. An hour later, seventeen of the patients were randomly selected to receive another placebo, while the other twenty-three patients were given naloxone, a drug that blocks endorphins (substances produced in the brain that reduce pain and whose chemical structure is similar to that of morphine). Surprisingly, the patients given naloxone reported much more pain than those given a second placebo. What was going on?

Levine and his co-workers proposed that naloxone was preventing the placebo response by blocking the morphine receptors in the brain, and that the production of endorphins was the mechanism by which placebos reduced pain. Since the publication of this seminal study, there has been an increasing body of evidence indicating that pain relief by placebos largely depends on their capacity to activate the brain's natural painkillers.

A recent brain imaging study conducted by neuroscientist Jon-Kar Zubieta and his co-workers at the University of Michigan–Ann Arbor received quite a bit of media attention and provided additional evidence that natural painkillers are involved in placebo analgesia.[17] These researchers used positron emission tomography (PET)—a nuclear medicine imaging technique that produces three-dimensional images of regional brain activity—to measure the activity of natural painkillers while sustained muscle pain was induced in young, healthy men by injecting salt water into the volunteers' jaws. During one of the PET scans, the investigators gave the volunteers a placebo infusion of salt water but told them that they were receiving a medication thought to have analgesic effects. The volunteers reported lower ratings of pain intensity. And compared with the condition in which no injection was given, PET scans showed that the salt water injection produced increased activity in brain regions known to inhibit pain through natural painkillers.

It would seem that the brain's natural painkillers are not the only factor in placebo response. In 2001, a team of scientists at the University of British Columbia in Vancouver made a discovery that provided further support for the view that different bodily processes appear to be implicated in distinct types of placebo responses. These scientists found a brain mechanism associated with the placebo effect in Parkinson's disease, a degenerative neurological disorder characterized by muscle rigidity, tremor, and a slowing of physical movement.[18]

The main problem in Parkinson's disease is a decline in the amount of dopamine, a chemical messenger, in brain structures called basal ganglia. The scientists took six Parkinsonian patients and scanned their brains with PET to compare the response to apomorphine (a drug that activates dopamine) and salt water. They administered both apomorphine and salt water in a double-blind fashion. Compared to those who received no treatment, patients who were given the placebo-induced expectation of motor improvement had both significantly improved motor performance and increased dopamine activity in the basal ganglia. Strikingly, the amount of dopamine activated in response to the placebo injection was comparable to that of the therapeutic dose of apomorphine given to the patients.

Stomach ulcers can also be affected by placebos. Anthropologist Daniel Moerman and his co-workers performed a meta-analysis[19]—a statistical approach that combines the results of several studies to test certain hypotheses in a more accurate way—merging the results of seventy-one controlled trials of drugs for treating stomach ulcers. In all the studies, an endoscope—a pencil-thin instrument used to examine the interior of a hollow organ—was used to determine whether the ulcers had really healed. Moerman and his colleagues compared those studies in which patients received two placebos a day to those in which patients received four placebos a day. The percentage of patients healed was significantly higher in the second group compared to the first group.

IIII

Emotional disorders are also candidates for the placebo effect. In one study, a group of researchers based at the University of Newcastle in the United Kingdom explored the effect of tablet color in the treatment of anxiety disorders.[20] Forty-eight patients who had been diagnosed as suffering from anxiety disorders were divided into three groups. All patients were given a mild tranquilizers, but the pills administered to each group were dyed with different colors—red, yellow, or green. During the second and third weeks, the colors were switched around so that eventually each group of patients tried each color of pills. Results revealed that the color of the pills had some impact on the effect of the medication. Specifically, green pills were the most efficient in reducing anxiety, and yellow the least efficient—even if the pills contained exactly the same amount of tranquilizer.

Meta-analyses have also convincingly demonstrated that placebo effects play an important role in clinical trials of mood disorders such as major depression. In 1998, psychologists Irving Kirsch and Guy Sapirstein examined the data for more than 2,000 depressed patients who had been randomly assigned to either antidepressant medications or placebo in nineteen double-blind clinical studies. Kirsch and Sapirstein found that 75 percent of therapeutic results were attributable to the placebo effect. That is, the inactive placebos produced improvement that represented 75 percent of the effect of the active drugs.[21]

||||

Clearly, the placebo effect plays a crucial role in treatments involving pills or injections. But there is mounting evidence that the placebo effect is critically involved in surgical interventions. Placebo (or sham) surgery is a fascinating insight into the power of mind over body.

In 1939, Italian surgeon Davide Fieschi introduced a new surgical procedure for treating angina, a chest pain due to lack of blood and oxygen getting to the heart muscle, usually related to the obstruction of the coronary arteries. Predicting that increased blood flow to the heart would decrease his patients' pain, Fieschi made small incisions in their chests and blocked two internal mammary arteries so that more blood and oxygen would become available to the heart. The results were striking: three-quarters of the patients showed improvement, and one-quarter of the patients were cured. Given its success, this technique became standard procedure for the treatment of angina for the next twenty years.

In 1959, American surgeon Leonard Cobb expressed doubts about this technique and decided to put the Fieschi procedure to the test. In a group of eight patients, he performed the standard Fieschi technique; in a second group of nine patients, Cobb made the incisions but did nothing more than that, letting the patients believe that they had received the real operation. Similar improvements were noted in both groups. This study, which led surgeons to abandon the Fieschi technique, marked the beginning of the documented surgical placebo effect.[22]

Surgical trials are important because they can reveal that a given surgical procedure is not better than a placebo procedure. In so doing, they can save thousands of patients from unnecessary surgery and postoperative pain.

In the 1990s, surgeon J. Bruce Moseley at the Baylor College of Medicine in Houston conducted a small placebo surgery trial of arthroscopy—a surgical procedure in which treatment of damage of the interior of a joint is performed—for osteoarthritis (a joint disease that causes the cartilage between the bones to wear away) of the knee.[23] Moseley recruited ten patients. The standard operation was given to two of them. Three patients received arthroscopic lavage—which flushes out the bad cartilage from the healthier tissue—without scraping, and five patients received a small incision with no lavage or scraping. Patients and medical researchers were blinded to the treatment group assignment. Moseley did not tell the patients whether they had received the real operation or the placebo.

All of the patients reported a significant decrease in pain. Members of the placebo group were able to walk and play basketball after the fake surgery. This is simply remarkable, given that some of these individuals walked with canes before the placebo intervention.

Surprised by the results, Moseley decided to perform another randomized, placebo-controlled trial to reach more definitive conclusions.[24] This time, 180 patients with osteoarthritis of the knee were recruited. The results were replicated: the outcomes after arthroscopic lavage or arthroscopic scraping were no better than those after a placebo procedure.

Another impressive surgical trial was performed in 2004 by Dr. Cynthia McRae and her colleagues at the University of Denver.[25] This double-blind sham surgery–controlled trial was designed to determine the effectiveness of transplantation of human embryonic dopamine neurons into the brains of individuals with advanced Parkinson's disease. The participants' quality of life was assessed one year after perceived treatment (that is, the type of surgery patients *thought* they had received). Participants recruited from across the United States and Canada were randomly assigned to receive the fetal tissue. Twelve patients received the transplant and eighteen received placebo surgery. Human embryonic dopamine neurons were implanted through tiny holes in the patients' brains. Needles for participants in the placebo condition remained empty and did not penetrate the brain. Regardless of which surgery patients had received, those who thought they received the transplant reported better quality of life a year later than those who thought they had received the placebo surgery. One patient in McRae's study reported that she had not been physically active for several years before surgery. In the year following surgery she resumed ice skating and hiking. When the double blind was lifted, she was amazed to discover that she had received the placebo surgery.

In the report, published in the *Archives of General Psychiatry,* Dr. McRae wrote that "medical staff, who did not know which treatment each patient received, also reported more differences and changes at twelve months based on patients' perceived treatment than on actual treatment." McRae also noted that although the placebo surgery research design is still somewhat controversial because of ethical concerns; the results of this study demonstrated "the importance of a double-blind design to distinguish the actual and perceived values of a treatment intervention."[26]

The results of the studies performed by Cobb, Moseley, and McRae suggest that placebo surgery can trigger an innate self-healing capacity. This capacity seemingly can exert a beneficial influence on a wide range of medical conditions and bodily functions.

Unfortunately, the administration of an inert placebo does not always elicit this self-healing capacity. It can sometimes lead to unpleasant and undesirable symptoms. This is called the *nocebo* effect, and it can happen when our expectations of treatment are negative rather than positive.

||||

The nocebo effect can follow negative expectations about one's own health. There is, for example, evidence that the belief that one is susceptible to heart attacks is itself a risk factor for coronary death. Epidemiologist Elaine Eaker has shown that women (forty-five to sixty-four years of age) who believe they are more likely than others to suffer a heart attack are nearly four times as likely to die of coronary conditions as are women who believe they are less likely to die of such symptoms (independent of other risk factors).[27]

The term *nocebo* (Latin for "I will harm") can also be used whenever symptom aggravation follows negative beliefs and expectations without the administration of any inert treatment. This term was originally invented to distinguish the detrimental or distressing effects of a placebo from its beneficial, therapeutic effects.[28] As with placebo effects, nocebo effects are influenced by the patient's perception of the treatment.[29] Roy Reeves and his colleagues at the University of Mississippi School of Medicine recently published a case report about Mr. A, a twenty-six-year-old man who experienced nocebo effects while taking a placebo during a drug trial.[30]

Mr. A had been depressed for a few months after his girlfriend broke up with him. When he saw an advertisement related to a clinical trial for a new antidepressant drug, he decided to enroll. During the first month of the trial, he felt that his mood improved considerably and he did not experience any problem with the capsules.

He had just begun his second month in the trial when he had an argument with his former girlfriend and attempted suicide by swallowing twenty-nine of the capsules. Suddenly remorseful and fearful that he would die of an overdose, Mr. A asked his neighbor to take him to the hospital.

He arrived at the hospital shaking and pale. His blood pressure was low and he was breathing rapidly. He was injected with normal saline to maintain an adequate arterial tension, and his blood pressure rose; but it dropped again when the infusion was slowed. Lethargic, he received nearly two hundred ounces of saline over a period of four hours. At this point, a doctor from the clinical trial arrived and told Mr. A that he had taken a placebo—twenty-nine of them. Mr. A expressed surprise followed by almost tearful relief. Within fifteen minutes, he was fully alert, and his blood pressure and heart rate were normal.

Like Mr. A, patients who expect distressing side effects before taking a medication are more likely to experience them afterward.[31] By now, it should also come as no surprise that the information patients receive about a drug changes their expectations of it and therefore their response to it. For instance, in a multicenter, placebo-controlled study of aspirin treatment for angina, the consent form at two of the participating centers listed "gastrointestinal irritation" as a potential side effect, but the consent form at the third center did not.[32] Patients at the first two institutions reported a significantly higher incidence of gastrointestinal symptoms than did the patients whose consent forms did not mention these side effects, and six times as many patients in the first group withdrew from the trial because of gastrointestinal distress. Along the same lines, among patients who received a muscle relaxant, those who were told it was a stimulant reported greater muscle tension than those who were told it was a relaxant.[33]

About 25 percent of patients taking placebos report negative side effects.[34] In placebo-controlled trials for antihypertensive medications and drugs to treat insufficient blood supply to the brain, side effect rates among those taking placebos are similar to those taking active drugs, and headache in particular is more frequent among those taking placebos.[35] Drowsiness, nausea, fatigue, and insomnia are the most commonly reported nocebo symptoms.[36]

A few studies have been conducted on brain activity associated with negative expectations; these have been performed essentially in the field of pain. Overall, negative expectations often result in the amplification of pain,[37] and several pain-related brain regions are activated during the anticipation of pain. Functional magnetic resonance imaging (fMRI) is an imaging technique that measures the changes in blood oxygenation related to neural activity in the brain. An fMRI study carried out by researchers at Wake Forest University School of Medicine demonstrated that as the degree of expected pain grows, activation increases in brain regions involved in pain perception. By contrast, expectations of reduced pain decreases activation of pain-related areas.[38]

||||

As we saw with Mr. A's experience, the impact of nocebo can be extremely important. Herbert Benson, an associate professor of medicine at the Mind/Body Medical Institute at Harvard Medical School, has proposed that "hex death" (also called "voodoo death") may represent an extreme form of the nocebo phenomenon.[39] Hex death is a death that follows a ritualized pronouncement of death by a person, such as a witch, sorcerer, or shaman, perceived by the hexed individual to have great power and authority. When death follows, it is attributed by the local community to the words and actions of the hexing individual.[40]

In one of his articles, Benson refers to the work of anthropologist Herbert Basedow, who in 1925 wrote one of the earliest descriptions of hex death in Western literature. Specifically, he reports the case of a member of an Australian aborigine tribe hexed by a shaman pointing a bone at him:

The man who discovers that he is being "boned" by any enemy is, indeed, a pitiable sight. He stands aghast, with his eyes staring at the treacherous pointer, and with his hands lifted as though to ward off the lethal medium, which he imagines is pouring into his body. His cheeks blanch and his eyes become glassy and the expression of his face becomes horribly distorted. . . . He attempts to shriek but usually the sound chokes in his throat, and all that one might see is froth at his mouth. His body begins to tremble and the muscles twist involuntarily. He sways backwards and falls to the ground, and after a short time appears to be in a swoon; but soon after he writhes as if in mortal agony, and, covering his face with his hands, begins to moan.

After a while he becomes very composed and crawls to his wurley. *From this time onwards he sickens and frets, refusing to eat and keeping aloof from the daily affairs of the tribe. Unless help is forthcoming in the shape of a countercharm administered by the hands of the* Nangarri, *or medicine-man, his death is only a matter of a comparatively short time.*[41]

Such extreme cases do not appear to be limited to remote tribes from traditional cultures. Clifton K. Meador, M.D., professor at the Vanderbilt University School of Medicine in Nashville, Tennessee, described two fascinating cases involving Americans.[42] The first case was reported to Meador by Drayton Daugherty, M.D., in 1961, and was confirmed in detail by a nurse and another doctor who witnessed the events.

In 1938, Dr. Daugherty admitted a sixty-year-old African American man (whom he calls "Vance Vanders") into a local hospital. Vanders had been ill for several weeks and had lost a lot of weight. He looked haggard and near death. He refused to eat and continued to decline despite a feeding tube, saying continuously that he was going to die. Soon he reached a state of near stupor and was hardly able to talk. The doctors were unable to identify a cause for his symptoms and stop his deterioration. At that point, Vanders's wife asked to talk to Dr. Daugherty privately.

The wife told Daugherty that about four months before being hospitalized, her husband had had a quarrel with a local voodoo priest. This priest had summoned Vanders to the cemetery late one night. In the midst of the quarrel, the priest waved a bottle of displeasing-smelling liquid in front of Vanders's face, and told him he had been "voodooed." The priest also announced that Vanders was about to die and that no one could do anything to save him. Back home, Vanders went to bed and began to decline.

Terrified, neither Vanders nor his wife had previously told the story: to keep them silent the voodoo priest had threatened to voodoo all their children. Dr. Daugherty spent many hours that evening mulling over the puzzling story and evaluating what he could do to save the moribund man.

The next morning, Daugherty called Vanders's family to his bedside. He told them authoritatively that the night before, he had enticed the voodoo priest to join him in the cemetery. There, Daugherty had physically forced the priest to explain how the curse worked. The voodoo priest had, Daugherty said, made Vanders breathe in some lizard eggs and they had climbed down into his stomach and hatched out small lizards. Daugherty further told Vanders and the members of his family that one lizard remained that was devouring Vanders from the inside out.

Daugherty also mentioned that he was now able to cure him of the terrible curse.

Drayton Daugherty then summoned a nurse who had, by previous agreement, filled a big syringe with a potent purgative. Ceremoniously, he injected its contents into Vanders's arm. Vanders began to vomit considerably a few minutes later. While Vanders was still vomiting, Daugherty secretly pulled a live green lizard from his bag and held it up. In a loud voice, he said, "Look what has come out of you! You are now cured. The voodoo curse is lifted."

Vanders's eyes widened and his mouth fell open. Flabbergasted, he quickly fell into a profound sleep. He woke up the following day, very hungry, and ate large quantities of food. He quickly recovered his strength and was discharged from the hospital a week later.

The other case presented by Dr. Clifton Meador acts as a cautionary tale. Meador was asked by a surgeon to see a hospitalized patient he calls "Sam Shoeman," who was suffering from a cancer of the esophagus. The small unshaven man in his seventies looked nearly dead when Meador saw him for the first time in October 1973. The surgeon told Meador that Shoeman's liver scan was quite abnormal, with evidence of carcinoma (a form of cancer) located in the entire left portion of the liver. Meador also had the operative report of the surgeon; hence he assumed that Shoeman was in need only of palliative care, because cancer of the esophagus was then incurable. Shoeman, the surgeon, and Shoeman's wife were all convinced that the cancer was in its terminal phase.

As expected, Sam Shoeman died a few months later, in January 1974. Yet the autopsy revealed that the doctors were mistaken:

only a small patch of bronchopneumonia was found, insufficient to cause death, as well as a tiny cancer nodule in the left lobe of the liver. The abnormal liver scan was a false-positive result, and the area around the esophagus was free of disease. "He didn't die from cancer, but from believing he was dying of cancer," said Meador. "If everyone treats you as if you are dying, you buy into it. Everything in your whole being becomes about dying." [43]

According to Meador, neither Vance Vanders nor Sam Shoeman had enough demonstrable organic disease to account for their dying state. Their cases suggest that patients who suffer noxious side effects may do so in part because of what their doctors told them to expect. [44]

||||

Today, high placebo responses are ruining several clinical drug trials and threatening the financial health of pharmaceutical giants, which in the 1990s were more profitable than major oil companies. The story of the antidepressant codenamed MK–869 illustrates well this alarming situation. In 2002, Merck's patents for several highly successful drugs were about to expire. In addition, the company's stock value was plunging. Merck had not generated a new drug in three years and was falling behind its rivals in sales.

Merck's research director, Edward Scolnick, proposed a plan to revitalize the company. A central element of his plan was to expand Merck's horizon into the antidepressant market, an area where the company had seriously lagged compared to competitors like GlaxoSmithKline and Pfizer. The plan depended largely on the success of MK–869. The first reports seemed encouraging. In clinical trials, the changes in brain chemistry induced by the drug appeared to foster well-being and revealed few side effects. Enthusiasm vanished quickly, however, once Scolnick and the Merck researchers realized that feelings of hopelessness and anxiety lifted in almost the same number of individuals taking MK–869 and a placebo (a pill made of milk sugar). Merck's attempt to enter the anti-depressant market had failed.[45]

In recent years, several other promising new drugs have been unable to beat sugar pills. From 2001 to 2006, the percentage of new drugs cut from development after Phase II clinical trials (when drugs are first tested against placebo) had risen by 20 percent. The FDA approved only nineteen new drugs in 2007—the lowest number since 1983—and twenty-four original drugs in 2008, despite all-time highs in R&D investment. In 2009, Osiris Therapeutics—a leading stem cell company—hit a setback when it was forced to suspend enrollment in a late-stage trial for its Crohn's disease treatment due to an unusually high placebo success rate in the trial. And a new type of gene therapy for Parkinson's disease, promoted by the Michael J. Fox Foundation, was suddenly withdrawn from Phase II trials after unexpectedly failing against placebo.[46]

Why does the placebo effect appear to be getting stronger? Some believe that the pharmaceutical industry's success in marketing its products may be part of the answer. In 1997, the FDA gave permission to Big Pharmas to run drug advertising campaigns. Since then, potential trial volunteers in the United States have been inundated with ads for prescription medications. In these ads, particular brand-name medications are associated with positive, uplifting aspects of life. Such associations may contribute to creating a huge placebo response.[47]

To contain the crisis, the Foundation for the National Institutes of Health (FNIH) has initiated the Placebo Response Drug Trials Survey. The main goal of this colossal study, which involves several decades of trial data, is to identify which factors are most responsible for the placebo effect. Major drug companies such as Lilly, GlaxoSmithKline, Pfizer, AstraZeneca, and Merck help fund the study.[48]

||||

Through our beliefs, we hold the power of life and death in our hands. Science has shown again and again that what we believe can significantly influence our experience of pain, the success of a surgery, even the outcome of a disease. Our expectations can trigger our bodies to do the work of regulating our physical and emotional experiences—a job that the Western world generally assigns to medications. And, as we have seen, they can even interfere with the efficacy of those drugs or with scientists' ability to judge whether those drugs are really doing the work.

Some researchers now conceptualize the placebo intervention as a set of signals that convey information. Examples of such signals include the treatment ritual, the therapist's white coat, the appearance of the therapist's office, and the words communicated by the therapist.[49] In addition, the way in which patients process these signals may be influenced by their history of experiences related to the milieu of therapy. The precise mechanisms that transform the detection of signals into placebo responses remain unknown. It seems likely that the signals associated with the placebo intervention are interpreted and translated into specific brain events. For instance, the expectation of a positive outcome may mobilize the neural networks involved in reward and lead to the production of dopamine.

IIII

The participants in the studies presented in this chapter were not intentionally inducing placebo and nocebo responses. But it is tempting to believe that we can deliberately alter our brain activity to improve our cognitive functions and emotional well-being. In the next chapter we explore this intriguing possibility.

2

Brain Control

Neurofeedback

Neurofeedback shows us how powerful we are.

—JOURNALIST JIM ROBBINS[1]

In his book *A Symphony in the Brain,*[2] science writer Jim Robbins recounts the poignant story of a boy named Jake who was born three months prematurely to a Helena, Montana, couple named Ray and Lisa. Jake weighed barely one pound at birth. When he was just three days old he underwent open-heart surgery and spent the next two months in intensive care. He survived, but he suffered brain damage.

At age four, Jake suddenly experienced his first of many grand mal seizures—a type of epileptic seizure that affects the entire brain—and became unconscious. Drugs diminished the seizures' severity but could not prevent them. Ray and Lisa often drove their son to the hospital emergency room, where all doctors could do was give him Valium injections to stop the convulsive attacks. Jake also suffered from petit mal seizures, during which his mind "turned off" and he could not hear or speak for several seconds.

As time went on, Jake's problems kept piling up: he was diagnosed with cerebral palsy, attention deficit disorder, hyperactivity, and speech problems. Moreover, his sleep was disturbed and he frequently woke up several times in the night.

At age five Jake began taking Depakote and Tegretol, two powerful anti-seizure drugs that produced troubling side effects, including fatigue and lethargy. His parents were worried, and Lisa described him as being "zoned out all the time."[3] Ray and Lisa began searching for alternatives to these medications. When they heard about neurofeedback, they felt this might give their son a chance to control his seizures on his own, without drugs.

Neurofeedback is a type of *biofeedback,* a re-training process in which individuals use real-time information about their body's responses (such as heart rhythm or muscular tension) to learn how to change aspects of their own physical functioning, and improve health and performance. Biofeedback instruments measure various kinds of physiological activity—including brain waves, heart function, breathing, muscle activity, and skin temperature—and rapidly feed this information back to the user. Using this information, while monitoring changes in how they are thinking and feeling, people can learn to produce at will the targeted physiological changes. Eventually, these changes can persist without the continued use of an instrument for feedback and monitoring.[4]

The goal of neurofeedback is specific: to learn to control the brain's electrical activity, described in terms of "waves" measured at cycles per second, or Hertz (Hz). These waves are measured with electroencephalography (EEG) using sensors (electrodes) attached to the scalp. These variations in brain activity may be slower or faster throughout the day, depending on various factors.

The slowest, delta waves (less than 4 Hz), are produced during sleep. Slightly faster, theta waves (4–7 Hz) occur with deep meditation. The so-called alpha state is an experience of relaxation that generates alpha waves of about 8–12 Hz. When we are solving problems, or sorting through issues, we produce beta waves (13–38 Hz). The fastest waves of all are gamma waves— 39–100 Hz. These waves are associated with higher mental activity.

After the electrodes are attached to the user's scalp (a painless, noninvasive procedure) the brain's activity is displayed to the user using sounds and images. Instantaneous changes in sounds and image—movement of a character in a video game, or a sound that decreases in pitch—let the user know how close he or she is to the desired target range of brain waves. If the change in the brain's electrical activity is not in the desired direction, no positive feedback is given. As users attempt to produce the desired sound or image, they are training themselves to control their brain waves.

When Ray and Lisa went on a search of neurofeedback training for Jake, the nearest site was located in a hospital in Jackson, Wyoming, three hundred miles away. They decided to make a series of appointments over the course of a week and make their son's treatment a family vacation. At Jackson's neurofeedback clinic, Jake, now eight, underwent two one-hour sessions every day that resembled games rather than the standard medical treatments he'd known his entire life.

Jake sat in the neurofeedback training room and stared attentively at a computer screen showing a Pacman that was beeping while gobbling dots. A small electrode attached to his scalp was connected to an EEG, recording his brain waves: now Jake's brain waves were directly influencing the Pacman, and Pacman's actions were giving Jake the feedback he needed to make adjustments in his brain waves. When Jake was able to produce the target frequencies by being more focused or breathing profoundly, the Pacman gobbled several dots and beeped a lot. When he was not in the target frequencies, the Pacman stopped gobbling and beeping. Jake quickly discovered how to adjust his brain waves in order to make the Pacman gobble dots and beep all the time, learning to calm himself in the process.

The results were profound. Following the week of neurofeedback training, Jake's sleep problems quickly disappeared. He was calmer and more focused, began to carry on conversations, and his motor skills greatly improved. Later, the neurofeedback protocol was repeated for another week. After this second week of training, Jake went to see his pediatric neurologist, Don Wright, who examined Jake and confirmed to Ray and Lisa what they already knew: the treatment had indeed been effective.

Jake's parents bought a neurofeedback machine and made it available to the people of Helena, and Don Wright decided to incorporate this technique into his clinical practice. Jake continued to train and improve. In 1999, he was evaluated for his individualized education program in the public schools. Lisa later told Robbins: "He was an emergent reader going into second grade and after a year of steady training, he was reading at a fourth-grade level. One of the teachers called Jake's rate of improvement explosive, and I think it was."[5]

Today, biofeedback and neurofeedback are commonly used techniques. But for a long time, researchers in physiology believed that humans could not have conscious control over brain activity. Fortunately, a few serendipitous discoveries during the second half of the twentieth century showed this assumption to be totally wrong.

IIII

Toward the end of the 1950s, University of Chicago psychology professor Joe Kamiya was wondering whether it would be possible for individuals continuously informed about their own brain waves to control them at will. Kamiya first designed an experiment to test whether a person could discriminate among his or her own brain wave categories and how he or she would describe the state. He focused on alpha waves because they are easy to produce. The participants in the experiment would lie in a dark chamber, eyes closed, and attached to an EEG machine. Kamiya then presented a sound, and asked participants by intercom to guess whether their brains were producing alpha waves. Using the EEG recordings, he could determine if the participants' guesses were correct and would respond "Correct" or "Wrong."

The first participant was Richard Bach, a graduate student. Kamiya placed an electrode to the left side of Bach's occiput (the back of the head), where alpha brain waves are naturally more abundant. In the first session, which consisted of sixty sounds and sixty guesses, Bach's guesses were about half correct. On the second day of testing, Bach reported correctly 65 percent. On the third day, he was 85 percent correct. Excited, Kamiya presented the sound four hundred times on the fourth day. Bach provided a correct answer each time.

In the second part of the experiment, Joe Kamiya asked Bach and other students to enter into the alpha state when a bell rang once, and not to enter into the alpha state when the bell rang twice. Quite a few students were able to enter and remain in the alpha state at will. This landmark experiment, which demonstrated that brain waves could be voluntarily controlled, initiated the field of neurofeedback.

Neurofeedback became widely known a decade later, when an article about this remarkable discovery was published in the popular magazine *Psychology Today*. In this article, Kamiya mentioned that some people reported feeling revivified and alert when they came out of the alpha state, and others reported experiencing feelings of tranquility and reverie or surges of creativity.[6]

University of California–Los Angeles (UCLA) neuroscientist Barry Sterman is another early pioneer in the field of neurofeedback. In 1965, he was investigating brain activity during internal inhibition—the process whereby a conditioned response is inhibited through lack of reinforcement. In one experiment, thirty cats were brought to his lab. Kept in a cage and deprived of food, these cats were trained to press a lever with a paw to receive a reward—a dose of chicken broth and milk. EEG electrodes were put on each cat's sensorimotor cortex, the part of the brain involved in sensory functions, and the control and execution of movements.

Once the cats were well conditioned to press the lever and obtain the reward, the conditioning procedure was modified. Now the cats had to wait until a sound stopped before they could press the lever and obtain the reward. The cats learned to remain completely still, yet very alert, while waiting for the sound to cease. Barry Sterman found that an unknown EEG rhythmic signal between 12 and 16 Hz was accompanying this motor still-ness. He dubbed this new EEG signal the *sensorimotor rhythm* (or SMR). Sterman naturally wondered whether a cat could be conditioned to produce SMR. For about a year, his assistants trained ten cats for an hour a day, three to four times a week. Sure enough, the cats learned to generate SMR.

Not long after, the U.S. Air Force asked Sterman if he wanted to test the negative cognitive effects of exposure to monomethyl-hydrazine. This rocket propellant was known to induce epilep-tic seizures and was thought to be affecting workers who were producing the substance, and even seemed to be affecting astro-nauts. Sterman accepted the Air Force's proposition. He injected the chemical into fifty cats. After an hour, most of these cats went into seizures. Three cats, however, did not suffer any sei-zure. Sterman realized that these three cats had been involved in his previous experiment and had learned to produce SMR. He hypothesized that the SMR training had enhanced resistance in the motor cortex of these cats against the slow theta waves re-sponsible for triggering seizures.

Sterman's next step was to explore whether SMR could be found in human beings. EEG recordings performed in patients who, because of cancer, had a part of the skull removed, con-firmed the existence of this rhythm in humans. Sterman decided to test the idea that people should be able to produce SMR.

He instructed his lab technician, Sid Ross, to build a neuro-feedback machine—a simple electronic box with two lights on it, one red and the other green—that he used with his first human subject, in 1972, twenty-three-year-old Mary Fairbanks. Since the age of eight, she had suffered from severe grand mal seizures two or more times every month. Sterman found that when Fairbanks was producing SMR and inhibiting low-frequency waves conducive to seizures, the green light came on. When she was not in the SMR range or was unable to block the slow waves, the red light appeared. Sterman asked Fairbanks to keep the green light on and the red light off as much as possible. She trained for an hour a day, twice a week, over the course of three months. At the end of the training, she was virtually seizure-free.

Sterman wrote a scientific article about Mary's experience.[7] In 1976 he received a grant from the Neurological Disease and Stroke Branch of the National Institutes of Health (NIH) to conduct a pilot study aimed at demonstrating the efficacy of the SMR training protocol. This study was based on an A-B-A design.

Eight epileptic patients were trained for three months to augment their SMR waves and inhibit low frequency waves. As expected, the number of seizures decreased markedly. (This was the first A segment of the study.) After three months the protocol was changed: the patients—who were not told what they were doing—were taught to increase low-frequency waves and decrease SMR. Unsurprisingly, they started experiencing seizures more frequently (the B part of the study). Three months later the protocol was modified again. Now, as in the first phase of the study, the patients had to reduce the frequency of their seizures by enhancing their SMR. Again the frequency of seizures diminished significantly. (This is the second A.) These impressive results were published in the *Epilepsia* journal in 1978.[8] A few years later, thanks to another NIH grant, Sterman was able to reproduce his findings, this time in twenty-four patients.

Since the 1970s, Sterman's pioneering work has been replicated in several other laboratories. During the past decade, two independent meta-analyses have been performed to evaluate the impact of SMR training in epilepsy. Overall, these meta-analyses included eighty-seven studies. Their results indicate that the SMR protocol leads to a significant reduction in seizure frequency in approximately 80 percent of the epileptics undergoing this type of neurofeedback training even when anti-seizure drugs are not working.[9]

||||

Studies indicate that approximately five percent of children are affected by attention deficit disorder (also known as ADD) and attention deficit hyperactivity disorder (or ADHD),[10] developmental problems that commonly appear during childhood. Children with these disorders are inattentive, impulsive, and in the case of ADHD, hyperactive. They have trouble sitting still and paying attention to one thing for a prolonged period of time. ADD and ADHD negatively affect academic performance and lead to increased risk for antisocial disorders and drug abuse in adulthood.[11]

ADHD children have high amounts of theta waves in the frontal lobe—a part of the brain that plays an important role in organizing behavior and controlling emotions. Some researchers believe that this excess of slow waves prevents the frontal lobe from communicating effectively with other cerebral structures.

In 1972, Joel Lubar, a research psychologist at the University of Tennessee in Knoxville, had already been investigating ADD/ADHD for several years. When he came across the first paper published by Barry Sterman about neurofeedback training in people with epilepsy, he immediately recognized a possible application of this work with people suffering from ADD/ ADHD. In 1976, Lubar moved to Los Angeles to work with Sterman for one year.

In his preliminary investigation, Lubar applied the Sterman's protocol to four children diagnosed with ADHD. He also used an A-B-A design, as Sterman did. The children were trained until psychological tests showed that their symptoms had vanished. Then they were trained in the reverse way until the tests demonstrated that their symptoms were back. Finally, the children were trained again until the tests and the EEGs confirmed that the symptoms were not present anymore. The protocol worked superbly and Lubar's insight was confirmed.

Since this early work, Lubar has conducted more than twenty-five studies regarding the effect of neurofeedback training in individuals with ADD or ADHD, and many studies have been carried out by other researchers. A meta-analysis published in 2009 indicates that the effects of neurofeedback in the treatment of ADD and ADHD can be regarded as clinically efficacious.[12]

Russell Barkley is a professor of psychiatry at the Medical University of South Carolina and an internationally recognized authority on ADHD. He has carried out several studies for drug companies and is an outspoken advocate of the use of psychostimulant drugs, such as Ritalin, to treat ADHD. Barkley claims that there is little evidence that neurofeedback works at all. He believes that a high placebo effect may explain why ADHD children get better following neurofeedback training. Barry Sterman agrees that a part of the results produced by neurofeedback training may be related to the placebo effect. However, this effect usually lasts only a short time, whereas clinical studies show that the results of neurofeedback tend to be long lasting.[13]

We do not, as yet, completely understand the neural mechanisms that underlie the effect of neurofeedback training. To explore this question, Johanne Lévesque (who was a postdoctoral fellow in my lab) used fMRI to measure the impact of neurofeedback training on brain regions implicated in selective attention (intentional, focused attention) and response inhibition (the suppression of an action that is inappropriate in a given context).[14] She recruited twenty children with ADHD, aged eight to twelve years. Fifteen children were randomly assigned to an experimental group, and received neurofeedback training. The other five children were assigned to a control group and did not undergo neurofeedback. All of the children were scanned one week before and one week after the end of training. While they were scanned, the children performed tasks that measure selective attention and response inhibition.

Neurofeedback training significantly improved performance on these tasks and decreased inattention and hyperactivity. This training also markedly enhanced the activation of brain areas involved in focused attention and the capacity to inhibit a response. No such changes were noted in the members of the control group. These findings indicate that neurofeedback training can improve the functioning of the brain regions implicated in attention and motor control.[15] In other words, *neurofeedback can functionally reorganize the brain*. It is plausible that such a functional reorganization is mediated by a strengthening of existing connections between neurons. It is also conceivable that new neuronal connections are created.

About this issue, Jimmy Ghaziri, one of my graduate students, recently found that the density of the white matter pathways linking brain regions implicated in attention is increased following a neurofeedback training protocol aimed at improving attentional performance in university students. This suggests that neurofeedback can reinforce neuronal connections between areas of the brain involved in cognitive functions. No white matter change was noted in participants who received sham neurofeedback.

||||

At the end of the 1980s, clinical psychologist Eugene Peniston and research psychologist Paul Kulkosky carried out a neurofeedback study at the Fort Lyon, Colorado, Veterans Hospital. Thirty men participated in this study. Twenty were alcoholic veterans who were back in the hospital for another round of in-patient treatment for alcoholism.

These patients were randomly divided into two groups. Ten received talk therapy and took part in a twelve-step process. Ten others received the same treatments and, in addition, the neurofeedback protocol. During training, the men who received neurofeedback were lying back in a recliner with eyes closed. They were instructed to allow sounds and the voice of a therapist to guide them into a deeply relaxed state associated with alpha and theta waves produced in the occipital cortex, the part of the brain lying in the back of the head.[16] They were also asked to use positive mental imagery (for example, being sober, refusing offers of alcohol, living confidently and happy) as they moved toward the trancelike alpha-theta state. Ten non-alcoholic men served as a control group.

The three groups of participants were administered EEGs and psychological tests before and after the twenty-eight-day treatment. The results in the patients receiving the alpha-theta protocol were striking. Their EEGs showed a considerable enhancement in occipital alpha and theta waves. This change in brain activity indicated that these patients were less anxious. Furthermore, the scores on the psychological tests revealed positive personality changes, and a marked decrease in negative emotional states. Eight out of the ten alcoholics who received the Peniston-Kulkosky protocol stopped drinking, and the ten patients who received traditional treatments were re-hospitalized within eighteen months. The abstinence lasted in the neurofeedback group. Three years later, only one participant had relapsed. These statistics are exceptional; in the field of substance abuse, a relapse rate of 70 to 80 percent is considered the norm.[17]

A few years later, Peniston used a similar approach with Vietnam War veterans who were suffering from posttraumatic stress disorder (PTSD). PTSD is an anxiety disorder that can develop after exposure to a frightening event in which serious physical harm occurred or was threatened. People with PTSD have persistent terrifying thoughts and memories that are responsible for nightmares, flashbacks, panic attacks, phobias, anxiety, and depression.

Two groups of veterans were included in this study. In one group, fourteen vets received traditional treatments—including psychotropic medications, individual therapy, and group therapy. In the other group, fifteen different vets received the alpha-theta training in addition to the traditional treatments. At the end of the study, nightmares, flashbacks, and the amount of psychotropic drugs had significantly diminished in the members of the neurofeedback group. No such changes were noted in the control group. A follow-up study was conducted thirty months later.

All fourteen vets in the control group had relapsed, while twelve of the fifteen vets who had received the Peniston-Kulkosky protocol were now living normally.[18]

Peniston has proposed that during an alpha-theta session, the veterans' peaceful physiology allows traumatic memories to smoothly emerge into awareness. He also speculated that in the theta state they feel like detached observers, so they do not need to reexperience the painful memories. This neutral mode of being would thus allow traumatic events to become painlessly integrated into the psyche.

IIII

As these neurofeedback studies show, it is relatively easy to learn to control our brain waves measured with EEG. But this technique has poor spatial resolution—that is, it does not allow researchers to precisely localize the brain regions that produce the electrical activity recorded at the scalp level. Given this limitation, it is not possible to determine with certainty whether one can learn to control activity in a particular brain region using EEG.

During the last decade, however, progress in neuroimaging technology has led to the development of the real-time functional MRI (rtfMRI). FMRI measures changes in blood flow and blood oxygenation that are closely related to the activity of neurons. The advantage of rtfMRI is that data are analyzed as they are collected. This small computation time allows scientists to quickly provide scanned individuals with visual feedback of ongoing activity in specific brain regions.[19]

Several rtfMRI neurofeedback studies have been conducted in recent years. In these studies, participants have to learn to control activity in a given brain region by employing the type of mental activity (thoughts, emotional feelings) that will maximize or minimize its activation. In one of these studies, researcher Christopher deCharms and his colleagues sought to determine the degree to which healthy individuals can learn to control activity in the somatomotor cortex, the motor portion of the brain.[20]

Participants were asked to imagine moving their dominant hand as they saw an image analog of the current level of activity in the somatomotor cortex. They were also instructed to increase activity in this region. Through training, participants succeeded in specifically enhancing their control over brain activity that was anatomically specific in the somatomotor cortex. Following training, using motor imagery alone, participants could increase at will activity in this brain region that was comparable in magnitude to the somatomotor activation measured during actual movement of the dominant hand. Notably, the participants were able to maintain their control over somatomotor activity even when real-time fMRI information was no longer provided.

In another study, deCharms and his co-workers used the same approach to train participants to control the activity levels in the rostral anterior cingulate cortex (rACC), a region of the brain known to be involved in the perception of pain.[21] The participants had to learn to increase (up-regulate) and decrease (down-regulate) the activity of the rACC while they received painful thermal stimulation. Remarkably, successful up-regulation of rACC activity—while the painful thermal stimulus was administered—resulted in subjectively higher reports of pain; in contrast, effective down-regulation of rACC activity during painful stimulation culminated in lower ratings of pain.

Other rtfMRI neurofeedback studies have shown that healthy people can quickly learn to control brain regions implicated in visual perception and hearing. It is possible that rtfMRI-based neurofeedback training might eventually be used to enhance performance—to stimulate activity in brain areas involved in memory, for example. In the future, rtfMRI-based neurofeedback might also be applied to anxiety and mood disorders.

J. Paul Hamilton, a neuroscientist at Stanford University, and his colleagues recently investigated whether individuals could use rtfMRI neurofeedback to learn to control the activity of the subgenual anterior cingulate cortex (sACC).[22] This other subdivision of the anterior cingulate cortex is thought to be involved in the production of emotional states, and has been implicated in major depression. In their study, Hamilton and co-workers asked eight women to down-regulate sACC activity by increasing positive mood. These women succeeded in diminishing activity in the sACC. This finding raises the possibility that cerebral structures that function abnormally in mood disorders can be controlled with the help of rtfMRI neurofeedback.

But neurofeedback is just one application of BCIs—brain-computer interfaces.

||||

Matt Nagle was paralyzed from the neck down as the result of a stabbing. In 2005, he became the first person to control an artificial hand using a BCI. A 96-electrode implant had been placed on the surface of the motor portion of his brain, over the area associated with his dominant left hand and arm. This implant also allowed Nagle to mentally control TV and check e-mails.[23]

A BCI first detects changes in brain signals that reflect the intention of the user and then translates these changes in signals into commands that carry out the desired action. Their main objective is to restore impaired or abolished movement, sight, and hearing. These communication pathways link the mind and brain of the users with their environments, allowing them to learn to control external devices, such as word-processing programs, switches, wheelchairs, televisions, and neuroprostheses.[24]

BCI systems can be driven by electrophysiological signals obtained from the scalp or directly within the brain. For example, a person can learn to use SMR activity by using various kinds of motor imagery to indicate "Yes" or "No" to control a cursor on a computer screen or a neuroprosthetic arm.[25] At the moment, BCI systems are beneficial mostly for people with major motor disabilities that prevent them from using voluntary muscle control. Such disabilities are seen in people with injuries to the spinal cord, amyotrophic lateral sclerosis (ALS, also known as Lou Gehrig's disease), severe cerebral palsy, muscular dystrophies, and acute disorders causing extensive paralysis (such as "locked-in syndrome").

Toward the end of the 1990s, German neuroscientist Niels Birbaumer and his colleagues developed a "mental typewriter" for ALS patients. These patients were trained to produce slow cortical potentials (SCPs)—negative or positive polarizations of the EEG—upon the command of an auditory cue. After achieving more than 70 percent control, the patients saw letters or words presented on a computer screen or heard them spoken by a word program. The patients were selecting letters by creating SCPs after the appearance of the desired letter.[26]

An increasing amount of BCI research is performed to allow humans to control their environment through their thoughts.

It is quite likely that noninvasive BCIs will soon be used by people who are not disabled. About this possibility, researchers at the IDIAP Research Institute, an organization specializing in the development of BCIs, recently reported a fascinating experiment. During the experiment, two healthy individuals succeeded in mentally moving a robot between several rooms, using an EEG-based BCI.[27]

Not long ago, Tzyy-Ping Jung, a researcher at the Swartz Center for Computational Neuroscience at the University of California–San Diego, and his colleagues developed a new BCI that could help severely disabled people communicate. This BCI allows users to place a call on a cell phone by thinking of the number—dialing with their thoughts. It is composed of an EEG headband that is connected to a Bluetooth module, which sends the EEG signals wirelessly to a Nokia cell phone. This BCI sounds like science fiction, but it is nearly 100 percent accurate for most users after only a short training period. Eventually, this *psychoneurophone* might be used by able-bodied cell-phone users.[28]

Working along the same lines, NeuroSky Inc., a venture company based in San Jose, California, has created a prototype that reads EEG brain waves using sensors attached to the user's forehead. This prototype, presented at the 2008 exhibition of the International Association for the Wireless Telecommunications Industry, displays the processed EEG signals on the screen of a mobile phone to show the degree of the user's relaxation. These signals can also be used to control the movement of a video game character who is shown on the screen of the mobile phone. NeuroSky is currently planning the development of BCIs that control home game consoles and home-use audiovisual equipment.[29]

Not surprisingly, mind control of brain activity has attracted the attention of the toy and games industry. A few years ago, the Mattel toy company marketed Mind Flex, a game that relies on NeuroSky's technology. Players wear a headset apparatus containing a forehead sensor that measures EEG activity. This activity is translated into a signal that is transmitted as a radio frequency. If the players concentrate hard enough, they can activate a fan that will make a ball rise and navigate through a tabletop obstacle course. The participants' ability to control their brain waves determines their success in guiding the ball through the course.

The Force Trainer is another game based on NeuroSky's brain-wave technology, this time a game that trains Jedi warriors. As with Mind Flex, a wireless headset reads the players' EEG activity. Players must get into a state of deep concentration and harness their brain waves to control a small ball moved by a flow of air inside a clear ten-inch-tall training tower. They can progress through several levels of training, from Padawan to Jedi, as they learn to use "The Force." Players are aided by instructions delivered by Yoda himself, the master of all Jedi Masters.[30]

Neurofeedback is efficient at reducing epileptic seizures, inattention, hyperactivity, and substance abuse.[31] And, as its use in gaming shows, this approach can also be used to optimize performance in healthy individuals. Because most amateur and professional athletes seek to sharpen their skills and enhance their performance, the world of sports is experiencing an increasing amount of interest in neurofeedback.

Vietta Wilson, a renowned sports psychologist and a professor at York University in Toronto, believes that the most central
element of athletic performance during competition is the control of mind. During the past three decades, she has trained athletes in many different sports, such as basketball, archery, track
and field, and wrestling. Her work has shown that worries and
negative thoughts are the worst enemies of athletic performance,
and that bio- and neurofeedback can help athletes to control
their mind and physiology.[32]

In 2002, sports psychologist Bruno Demichelis—who was
then head of Sports Science for the professional soccer team AC
Milan—studied under Vietta Wilson. Next, he created a "secret
weapon" that he called the *Mind Room*. In this room, the AC
Milan players lay down on reclining chairs, their bodies connected to biofeedback machines that were measuring their brain
waves, heart rate, and muscular tension. Demichelis taught them
how to promptly attain and remain in a relaxed state while they
were watching videos of their mistakes. The players eventually
learned to acquire mental and physiological control. In 2006,
some of these players helped the Italian team to win the World
Cup. The following year, AC Milan won the European cham-
pionship.[33]

The Los Angeles Clipper—seven-foot National Basketball
Association (NBA) center Chris Kaman—is another athlete
who benefited greatly from neurofeedback training. In his early
years in the NBA, he could not concentrate and often forgot
what he was doing. He was also very impulsive. Recently, he underwent a series of neurofeedback sessions under the supervision
of psychologist Tim Royer. Following this training he averaged
a career-high 17.9 points, 13.7 rebounds, and three blocks per

game, and became a dominant center in the NBA. Kaman attributes his athletic improvements to neurofeedback. Brain wave training helped him gain new abilities, including concentration and impulse control.[34]

Skier Alexandre Bilodeau also used bio- and neurofeedback to train his mind and body. He won the men's moguls Gold Medal at the 2010 Vancouver Winter Olympics. Bioneurofeedback was one of several projects labeled *Top Secret,* the science and technology component of *Own the Podium.* This $117-million, five-year plan was engineered to help Canada win the most medals at the last Olympics. Bilodeau was trained by Penny Werthner, a sports psychologist and a professor at the University of Ottawa.

Bioneurofeedback taught Bilodeau how to mentally switch into performance mode for 25 to 30 seconds down the hill, and relax between runs. "Focus takes a lot of energy," Werthner said. "It's a very difficult balance to be very intense, committed to do well, and yet have this calmness of 'I can do this.' That's an incredibly hard combination to get. It's not easy to win an Olympic medal and Alex was brilliant. I found this tool really useful—it's not some miracle thing by any means—but a useful way to help athletes become much more self-aware, but most important to train to change."[35]

Neurofeedback is a potent form of biofeedback that allows us to deliberately change what is going on in our brains. This psycho-neuro technology gives us a glimpse of the remarkable power of our minds. Neurofeedback can enhance our cognitive functions, reduce anxiety and mood disorders, and lead to greater emotional well-being. There is also some evidence indicating that certain types of brain wave training can contribute to the occurrence of transcendent experiences. Moreover, real-time fMRI neurofeedback studies show that we can learn to control the activity of specific brain regions, and BCIs demonstrate that we can influence our environment with our thoughts. For all these reasons, I am convinced that the invention of the neurofeedback technology marks a significant step in our evolution.

But neurofeedback is not the only way we can influence our brains. The next chapter explores recent neuroscience research demonstrating that meditation—a form of mental training almost as old as humankind—can also have a beneficial impact on brain activity. In fact, there is evidence to show that it can, quite literally, "change our brains."

3

Train Your Mind,
Transform Your Brain

Neuroplasticity

In a real sense the brain we develop reflects the life we lead.

—THE DALAI LAMA[1]

In 2004, the fourteenth Dalai Lama, Tenzin Gyatso, invited
five neuroscientists to Dharamsala, India. The main topic of
this gathering was neuroplasticity: the dynamic potential of the
brain to reorganize itself throughout life in response to everyday experience. Dozens of Tibetan Buddhist monks attended
this meeting, which lasted five days. The neuroscientists unani-
mously supported the notion that mind is only a manifestation
of electrical and chemical processes in the brain, and that there is
no need to have recourse to anything spiritual or nonphysical to

understand mental capacities. Expectedly, the Dalai Lama and the monks had a very different point of view.[2]

Despite their divergent beliefs, monks and scientists alike were enthusiastic about discussing the enticing possibility that mind shapes brain in foreseeable fashion, in much the same way as physical training reliably conditions muscles. The Dalai Lama expressed his profound conviction that thoughts and emotional feelings, while internal and intangible, can have a significant influence upon the activity and structure of the brain.

Neuroscientist Fred Gage, who attended the 2004 Dharamsala meeting, frames the traditional view of the brain in this way: "If the brain was changeable, then we would change. And if the brain made wrong changes, then we would change incorrectly. It was easier to believe there were no changes. That way, the individual would remain pretty much fixed."[3]

Not all researchers welcomed the Dalai Lama into neuroscience with open arms. In 2005, when the Society for Neuroscience (SfN)—the largest professional society in the world for neuroscientists—notified members that the Dalai Lama had agreed to be the first-ever speaker in an annual lecture series, "Dialogues Between Neuroscience and Society," at the upcoming fall meeting in Washington, D.C., it did not go unnoticed. Although his talk, "The Neuroscience of Meditation," was designed to provide an opportunity for the Tibetan leader to promote the idea of a partnership between Buddhism and science, it soon ran into strong opposition.

Some neuroscientists urged SfN to cancel the lecture, and threatened to boycott the meeting. The Dalai Lama, they complained, is not qualified to talk about neuroscience. A researcher at the National Institutes of Health said, "We don't want to mix science and religion in our children's classrooms, and we don't want it at a scientific meeting." Another petition organizer was even more trenchant, asking, "Who's coming next year? The Pope?"

||||

The history of research into the brain's capacity to change and grow with learning is not long. In the early 1960s Mark Rosenzweig, a research psychologist at the University of California–Berkeley, became one of the early pioneers of neuroplasticity. He and his fellow researchers found that rats raised in an enriched environment—cages filled with running wheels, toys to roll, and ladders to climb—were able to learn better than were genetically similar rats reared in an unenriched environment, bare of such objects. When they later examined the rats' brains, the results were clear: the brains of the rats raised in the enriched environment weighed more and contained more chemical messengers than those from the rats reared in the unenriched environment.[4]

About thirty years later, in the late 1990s, Fred Gage and his colleagues at the Salk Institute compared the brains of two sets of aging mice—littermates raised in enriched and unenriched environments. They showed that the mice raised in the enriched environment were superior on tests evaluating exploration and learning, and that their brains displayed significantly more new neurons in the hippocampus compared with the brains of those raised in the unenriched environment. This important discovery indicates that environmental stimulation can promote neurogenesis—the generation of new neurons—even in an aging brain.[5]

Until the 1970s, it had been a central dogma of neuroscience that the adult brain was a static "hardwired" machine, with no ability to change and produce new neurons. Yet the scientific studies that led to the revolutionary discovery of neuroplasticity and other landmark studies have shown just the reverse: the adult human brain is continually changing its structure and function by creating new neurons and synaptic connections, and reorganizing existing neuronal networks or elaborating novel networks. We are not stuck with the brains we were born with. Just ask a London cabbie.

Drivers of London's iconic black cabs can expect to earn almost twice as much as other cabbies—but it's an arduous road to that reward. Would-be cabbies must become intimately acquainted with the multitude of streets and individual neighborhoods that lay inside a six-mile radius of Charing Cross in central London, and pass an intimidating oral test called "The Knowledge." The study and training for the test is costly and takes drivers several years of hard work to complete. "The average student does 15 to 30 hours a week study, for three years," according to *The Observer.*[6] But once drivers earn the coveted license, they have done more than pass a test: they have changed the size of their brains.

In 2000, a research team at London's University College led by Dr. Eleanor Maguire conducted magnetic resonance imaging (MRI) brain scans of sixteen London taxi drivers who had extensive navigation experience. The researchers compared the brains of these taxi drivers to those of control subjects who did not drive taxis. Maguire and her colleagues found compelling evidence that the brains of adults can, indeed, be physically changed by knowledge.

In each hemisphere of the brain, the posterior hippocampi of the taxi drivers were significantly larger than those of control subjects. The posterior hippocampus, located in the medial temporal lobe of the brain, consolidates information from short-term to long-term memory, plays a part in spatial navigation, and is thought to store a spatial representation of the environment. It is not surprising that this region was found to be even more developed in taxi drivers who had been in the career for several decades than in those who had been driving for a shorter span of time. Still, even the drivers were surprised. "I never noticed part of my brain growing," said one. "It makes you wonder what happened to the rest of it."[7]

What these findings suggest, say Maguire and her colleagues, is that the posterior part of the hippocampus can expand in people with a high dependence on navigational skills. Our brains are not fixed; they can grow and change over time, depending on how we use them. This supports the view of many neuroscientists that there is a capacity for regional plastic change in the structure of the adult human brain in response to demands of the environment. "This is the first study to show that the work you do can really change the structure of the brain," said Maguire. "This insight into the plasticity of the human brain might offer hope for rehabilitation of neurologically injured patients."[8]

▋▋▋▋

It's not only acquired knowledge that has an impact on neuroplasticity. Research has also shown that changes in thoughts and feelings have the power to transform the brain.

My research team and I demonstrated this some years ago with a group of young women suffering from arachnophobia. This irrational fear of spiders can be so intense as to trigger panic attacks, even when a living spider is not actually present.

In our experiment, we asked these spider phobics to watch film excerpts of live spiders in motion while we scanned their brains with fMRI. All of the participants experienced intense fearful feelings as they watched the spiders on the screen, and the fMRI scans revealed that the fear reaction was associated with a strong activation of the hippocampal formation.

We know that phobias are characterized by phobic avoidance: if you are afraid of spiders, you will go to great lengths to get away from them. This impulse arises from an association of panic attacks with the context in which the fear reaction originally occurred. For example, if you became fearful of spiders because you encountered a nest of them inside a dark closet, the very act of opening a closet door may trigger a panic attack. The hippocampal formation plays an important role in the memory through which fear conditioning is established. Because the majority of the young women examined in our study developed a phobia following distressing childhood experiences with spiders, we proposed that the activation of the hippocampal formation was related to the emotional memories associated with these negative experiences.[9]

One week later, our phobics began a desensitization therapy designed to lessen their fear of spiders. The young women met for one three-hour group session each week for four weeks. The procedure was straightforward, gradually increasing education and exposure: The first week, they were asked to look at a book containing color pictures of spiders. The second week, they were shown film excerpts of living spiders, and were asked to look at the pictures and watch the film clips at home between sessions. The third week, they were asked to stay in a room that contained living spiders. During the fourth and last session, they were requested to touch a huge, live tarantula. All of the participants were able to do this successfully—quite remarkable, considering that before therapy, most were so phobic that they were unable to touch even *pictures* of spiders.

A week after the end of therapy, we again scanned the participants as they watched film excerpts of moving spiders. This time, the film excerpts did not produce fearful feelings, and the scans backed up their responses: they showed no activation of the hippocampal formation. These impressive findings suggest that the participants had functionally "rewired" their own brains, over a period of only a month, so that they no longer felt the fear that had restricted their lives.[10] Coincidentally, about a year after the end of the study, I met one of the participants. She told me that after the study she had become very much enamored of spiders—so much so, that she had just adopted a giant tarantula as a pet.

▮▮▮▮

Emotions can be very destructive; indeed, they constitute one of the main causes of human suffering. Fortunately, most of us are not at the mercy of our emotions but can modulate our emotional responses at will.

Back in 2001, I started a research program to explore what happens in the brain when healthy people are asked to take control of their emotions. In our first study, we asked ten young men to watch excerpts from erotic films. The participants were scanned with fMRI in a control condition and an experimental condition. In the control condition, they were instructed to simply watch the film excerpts and react normally. In the experimental condition, the men were requested to observe comparable but not identical film excerpts in a dispassionate, non-evaluative, and nonjudgmental way.

As expected, all of the participants were sexually aroused by the erotic film clips. In the control condition, sexual arousal was associated with activation in various brain structures known to be involved in emotions, such as the amygdala and the hypothalamus. The participants were all able to decrease their arousal in the experimental condition, and no activation was detected in these cerebral structures in response to the erotic videos.[11]

Later, my colleagues and I conducted another fMRI study using a similar approach.[12] This time, however, we measured the brain activity of twenty psychologically healthy young women attempting to control sad feelings evoked by film excerpts. These excerpts featured the death of a beloved person. The women reported that they were able to reduce the sad feelings in the experimental condition. We found that the reduction of sad feelings was also accompanied by decreased activation in brain areas implicated in sadness.

The results of these studies indicate that normal people are not "feeling machines" who simply respond to stimuli. They are very much capable of controlling their reactions and the responses of their brains to emotional events.

Can mentally healthy people also influence the activity of chemical messengers that play a role in emotions? To investigate this crucial question, we used positron emission tomography (PET) to estimate the production of serotonin during rapid and sustained changes of emotional state.[13] This chemical messenger is known to be crucially involved in mood and the control of emotion.

Seven healthy professional actors, all of them method actors, participated in our PET study. Method acting is a technique advocated by Lee Strasberg, famed director of the Actors Studio in New York City in the 1950s, and used by actors such as Marlon Brando, Dustin Hoffman, and many others. In this approach actors draw on their own emotions and emotional memories to power their portrayals. In contrast, more traditional forms of acting use techniques in which actors only simulate the emotions of their characters.

We asked the participating actors to self-induce transient states of sadness and happiness. To do so, they were instructed to relive and re-enact intense, genuine emotions associated with specific autobiographical memories. The actors underwent scanning sessions on separate days (one session for sadness, the other session for happiness). After each scanning session, they were asked to report the intensity of the emotions they experienced on a self-report scale.

All of the actors reported that they significantly experienced the target emotional states, and the PET results mirrored their subjective reports in distinct ways: the actors' reported levels of sadness were correlated with a reduction of serotonin production in the brain's emotional regions, such as the orbitofrontal cortex and the anterior cingulate cortex. In contrast, the intensity of the happy feelings was associated with increased serotonin production in emotional areas of the brain. These findings are consistent with the evidence indicating that serotonin activity is diminished in these areas in individuals with major depression.

The results of our brain imaging studies suggest that it is possible to rapidly influence brain chemicals related to emotions and mood, as well as the activity of the brain regions implicated in emotional reactions. If this is possible, then what results can mind and brain training have over the long term? A few curious neuroscientists, myself included, are currently on the road to find out.

||||

Most people who have tried meditation, even casually, agree that it has a calming effect. Many people who meditate regularly would say that in general, they feel more peaceful and are able to think more clearly. But is it truly possible to make real changes to the brain simply by meditating regularly? The Dalai Lama believes it is. "It is a fundamental Buddhist principle that the human mind has a tremendous potential for transformation," he has written. "Buddhist practitioners familiar with the workings of the mind have long been aware that it can be transformed through training. . . . In a real sense the brain we develop reflects the life we lead."[14]

Science backs him up. A number of neuroscientific studies performed in recent years have demonstrated that willful attention and its training through the practice of meditation can indeed lead to important plastic changes in the brain. The broad term *meditation* refers to a large variety of mental training techniques that have been developed for various purposes, including fostering emotional balance and well-being.[15] These techniques are generally classified into two types: mindfulness and concentrative. *Mindfulness* practices involve allowing any sensations, thoughts, or feelings to arise from moment to moment, while maintaining awareness as an attentive and nonattached observer without judgment or analysis. Such practices are part of ancient Eastern traditions of meditation such as Vipassana and Zen.[16]

Concentrative meditational practices involve focusing attention on specific body sensations (such as breath) and mental activity (such as a repeated sound or an imagined image). Examples include forms of yogic meditation and the Buddhist Shamatha meditation focus on the sensation of breath. Both mindfulness and concentrative techniques elicit a deep sense of calm peacefulness, a slowing of the mind's internal dialogue, and a shift toward an expanded experience of self not centered on the meditator's body representations and thoughts.[17]

Three decades ago, Jon Kabat-Zinn, a biomedical scientist at the University of Massachusetts Medical School and a student of Zen Master Seung Sahn, developed the Mindfulness-Based Stress Reduction (MBSR). MBSR is an eight-week intensive group program that brings together mindfulness meditation and yoga. Over the past thirty years, several thousands of people have taken the MBSR course, and studies have shown that MBSR significantly decreases stress symptoms in people with various types of cancer. Research has also demonstrated that MBSR reduces depression, rumination, and anxiety, and promotes well-being, compassion, and spirituality.

These potentially life-altering findings come as no surprise to longtime practitioners, including the Dalai Lama. Internationally recognized as a proponent of compassion, universal responsibility, and the nonviolent resolution of human conflict, he won the Nobel Prize for Peace in 1989. And he has always demonstrated a fervent interest in science, developing personal relationships with renowned physicists Carl von Weizsäcker and David Bohm and philosopher of science Karl Popper. The Tibetan leader has also participated in several conferences on science and spirituality. He is deeply convinced that science provides efficient means for understanding the basic interconnectedness of all life.

He believes that science and Buddhism should both contribute to a better comprehension of the world: an important rationale for ethical behavior and environmental protection. In keeping with this conception, the Dalai Lama has asked Tibetan scholars to become familiar with science in order to rekindle the Tibetan philosophical tradition.

At the Alpbach Symposium on Consciousness in 1983, the Dalai Lama met Francisco Varela, the late Chilean-born neuroscientist who had become a Tibetan Buddhist in the 1970s. During that meeting, they began a discussion about consciousness and neuroscience. Varela agreed with the Dalai Lama that Buddhism represents an important source of observations concerning the human mind, with specific mental techniques to improve cognitive function and emotional well-being. Following this symposium Varela, in collaboration with Adam Engle—a lawyer and businessman, and also a Buddhist practitioner—created a series of meetings about Buddhism and science. These meetings, in-depth dialogues between the Dalai Lama and Western philosophers and scientists, formed the basis of what became the Mind & Life Institute. The first Mind & Life meeting was held in 1987 in Dharamsala, in northern India, the home of the Dalai Lama and of the exiled Tibetan government since Chinese troops invaded Tibet.[18]

Beginning in the 1990s, the Dalai Lama has arranged for Tibetan Buddhist monks with considerable meditation experience (at least 10,000 hours of practice) to travel to American universities to participate in brain-imaging studies. These studies seek to investigate whether Buddhist meditative techniques can produce lasting changes in the brain. So in 2005, when some scientists protested the Dalai Lama's talk at the SfN conference, it came as something of a surprise.

A neuroscientist at the University of Florida, Jianguo Gu, helped to set up an online petition against the lecture: "I don't think it's appropriate to have a prominent religious leader at a scientific event," he explained. "The Dalai Lama basically says the body and mind can be separated and passed to other people. There are no scientific grounds for that. We'll be talking about cells and molecules and he's going to talk about something that isn't there." (Although Gu and several of the scientists who started the protest are of Chinese origin, they stated that they were not against Buddhism. Rather, their main concern was to avoid entanglement with religion or politics, and confusion between objective inquiry and faith.[19])

In the petition, the protestors asserted that

inviting the Dalai Lama to lecture on "Neuroscience of Meditation" is of poor scientific taste because it will highlight a subject with largely unsubstantiated claims and compromised scientific rigor and objectivity at a prestigious meeting attended by more than 20,000 neuroscientists. . . . It is ironic for neuroscientists to provide a forum for and, with it, implicit endorsement of a religious leader whose legitimacy relies on reincarnation, a doctrine against the very foundation of neuroscience. The present Dalai Lama explicitly claims the separation of mind and body, which is essential to the recognition of the Dalai Lama as both a religious and a political leader. It would serve the interests of SfN as well as the public to cancel the talk.[20]

Carol Barnes, the president of the SfN, responded, "The Dalai Lama has had a long interest in science and has maintained an ongoing dialogue with leading neuroscientists for more than fifteen years, which is the reason he was invited to speak at the meeting. It has been agreed that the talk will not be about religion or politics. We understand that not every member will agree with every decision and we respect their right to disagree."[21]

Most protesters effectively boycotted the meeting and withdrew their conference papers. Acclaimed neuroscientist and one of the Dalai Lama's primary scientific collaborators Richard Davidson was criticized by Yi Rao, a research neurologist at Northwestern University and a leader of the opponents to the Dalai Lama's lecture. Rao said, "The motivations of both Davidson and the Dalai Lama are questionable." He also accused Davidson of being a "politically involved scientist" who organized the Dalai Lama's speech to confer scientific legitimacy to Buddhism. Davidson answered back that the opposition to the speech was obviously driven by the Chinese government's long-running propaganda campaign against the Tibetan leader.

Davidson's close personal relationship with the Dalai Lama has been questioned on the basis that scientists are supposed to keep professional distance from organizations and individuals supporting their research projects. Davidson responds that he greatly values his relationship with the Dalai Lama, and has no intention of giving it up. Some have pointed out that researchers gravitating around the Mind & Life Institute are at risk of losing their objectivity and influencing the results of their experiments by becoming acolytes of the Dalai Lama. About this, Charles Raison, a research psychiatrist at Emory University who has investigated the impact of meditation on the immune system, notes, "This is a field that has been populated by true believers. Many of the people doing this research are trying to prove scientifically what they already know from experience, which is a major flaw." But Davidson contends that several scientists have profound personal interest in what they are studying, and this is a good thing.[22]

This was a contentious matter that threatened to upend the conference. And it also serves to reinforce an issue that we are investigating throughout this book: the tension between the restrictive dogmas of Western science and the free investigation of phenomena.

IIII

The notion of investigating what is going on in the brain during meditation is far from a new or even radical idea. In the 1950s pioneer researchers carried electroencephalography (EEG) devices up into the mountain caves of Indian yogis and conducted the first studies exploring brain activity during meditation. Since then, a number of EEG studies have been performed with various types of meditation techniques. These studies have revealed results concerning brain waves that have very interesting implications for all of us.

Research shows increases in alpha and theta activity during mindfulness meditative practices, such as Vipassana and Zen— not surprising, because increased alpha wave activity is thought to reflect relaxation, and theta waves are believed to be a specific marker of mindfulness meditative states. Some investigations of Zen meditation indicate that the magnitude of increases in theta activity is related to the level of expertise of the practitioners.[23]

Other EEG studies have shown that increased beta 2 (20 to 30 Hz) and gamma activity characterizes concentrative states of meditation. Beta 2 activity is reported during tasks involving focused attention; gamma activity is believed to be related to consciousness and the content of mental experience.[24]

Overall, the EEG studies of meditation conducted to date confirm the idea that different neuroelectrical signatures accompany different types of meditative practices. This makes a lot of sense, considering that different meditative practices are characterized by distinct mental processes and contents of experience.

In 2004, Richard Davidson, Antoine Lutz, and their colleagues at the University of Wisconsin–Madison published the results of an EEG study that received a lot of attention from the media.[25] In that study, they recruited eight very experienced Tibetan Buddhist monks and ten novices (healthy student volunteers) who had had an introductory course in meditation. While the participants engaged in a form of meditation called nonreferential compassion, researchers measured their brain waves. During this kind of meditative state, meditators are asked to focus on unlimited compassion and loving kindness toward all living beings. Before the study, the novices had practiced this form of meditation for only one week (one hour daily). The Tibetan Buddhist monks had practiced nonreferential compassion for periods ranging from fifteen to forty years.

The researchers recorded exceptionally large increases in gamma waves in the monks for the nonreferential compassion state, compared with a resting state. These gamma waves were much more intense in the Tibetan monks than in the novices. Remarkably, higher gamma activity was still seen in the monks when they stopped meditating. Moreover, the more hours of meditation training a monk had had, the more robust and lasting the gamma activity. These findings suggest that meditative training may induce long-term changes in brain activity, even outside of meditation.[26]

That study was criticized on the ground that age could have accounted for some of the differences found: the monks investigated were twelve to forty-five years older than the university students. In addition, there was no way to know whether the monks had been adept at producing high gamma wave activity before they ever began meditating. Nonetheless, the results of this study are intriguing.

In another study, the research team led by Richard Davidson used fMRI to examine what happens in the brain during "one-pointed concentration," a form of meditation that is practiced to increase attentional focus and reach a peaceful state in which preoccupation with thoughts and emotions is progressively diminished.[27] They compared a group of Tibetan Buddhist monks with extensive meditation experience to a group of age-matched novice meditators. Davidson and his colleagues found that activation in brain regions normally implicated in sustained attention was generally more robust for the expert meditators compared to novices. However, whereas the monks with an average of 19,000 hours of practice exhibited greater activation in these regions than the novices, those monks with an average of 44,000 practice hours showed *less* activation. This fits well with meditation texts that present concentration meditation as requiring at first higher levels of effortful concentration but eventually becoming less effortful, such that later phases of this meditative practice necessitate minor effort.

Davidson and his colleagues have also used fMRI to measure brain activity while Tibetan monks and novices voluntarily produced a loving-kindness and compassion meditation state.[28] These researchers saw greater activation of brain areas implicated in empathy in the monks, compared to the novices. This finding demonstrates that the mental expertise to self-induce positive emotion alters the activation of cerebral structures known to be involved in empathic responses.

Recently, Véronique Taylor, a master's student in my lab at the University of Montreal, conducted an fMRI study to explore the effects of mindfulness meditation on the brain responses to emotionally evocative color pictures.[29] Another goal of this study was to examine the impact of the duration of mindfulness training on the brain responses to such pictures. Experienced meditators with more than 1,000 hours of experience in Zen meditation were compared to novice meditators. Novices were instructed to practice mindfulness meditation twenty minutes per day for seven days before the experiment. The two groups of participants were scanned as they viewed negative, positive, and neutral pictures in a mindful state and a nonmindful state of awareness.

Both groups subjectively perceived the pictures viewed in a mindful state as less intense than when viewed in a nonmindful state. Furthermore, in experienced meditators, mindfulness was accompanied by decreased activation in brain regions typically associated with emotional reactivity. These results support the view that mindfulness meditation eases the impact of emotional triggers. Our results also suggest that with extensive training, mindfulness meditation may promote a state of mental calmness by quieting brain activity.

IIII

Can the practice of meditation also lead to changes in the actual structure of the brain? A few years ago, research psychologist Sara Lazar and her colleagues at Harvard University decided to tackle this important question using structural MRI. This method provides the most accurate information regarding the anatomy of the brain. Lazar and her co-workers showed that the long-term practice of meditation is indeed associated with changes in the brain's physical structure.

They compared anatomical brain scans of fifteen non-meditators to those of twenty experienced meditators who had an average of nearly 3,000 hours of mindfulness practice. Increased thickness of gray matter was found in brain regions associated with attention and interoception, the capacity to consciously detect changes occurring within the body.[30]

"Our data suggest that meditation practice can promote plasticity in adults in areas important for cognitive and emotional processing and well-being," said Lazar. "These findings are consistent with other studies that demonstrated increased thickness of music areas in the brains of musicians, and visual and motor areas in the brains of jugglers. In other words, the structure of an adult brain can change in response to repeated practice." Lazar also noted that the increases in gray matter thickness were proportional to the time the meditators had been practicing mindfulness.[31]

Britta Hölzel, another research psychologist at Harvard University, has conducted a structural MRI study to measure gray matter changes induced by the MBSR program.[32] Anatomical scans from sixteen meditation-naïve participants were obtained before and after they underwent MBSR. The MBSR group was compared with a control group of seventeen individuals who did not practice meditation. Participants in the MBSR group meditated for about forty-five minutes a day for eight weeks. In participants who received MBSR, increases in gray matter density were measured in brain regions implicated in learning and memory, empathy, and emotion regulation—that is, the ability to engage in healthy strategies to manage disagreeable emotions when necessary. No such changes were seen in the control group.

In yet another study, neuroscientists Giuseppe Pagnoni and Milos Cekic at Emory University in Atlanta, Georgia, have used structural MRI to examine how the regular practice of meditation may affect the normal age-related decline of cerebral gray matter volume and attentional performance seen in healthy people.[33] Gray matter volume has been shown to decrease immediately after adolescence.

Pagnoni and Cekic recruited regular practitioners of Zen meditation who had more than three years of daily practice, and control subjects who had never practiced meditation. The two groups were matched by gender, level of education, and age (thirty-seven years for the meditators, thirty-five years for the control subjects); they all participated in a task assessing vigilance, a sustained form of attention.

The expected negative correlation of gray matter volume and attentional performance with age was found in the control subjects; no such correlation was detected in the meditators. In addition, gray matter volume in the putamen, a cerebral structure involved in attention, was positively correlated with performance on the vigilance task. These findings suggest that the regular practice of Zen meditation may offer protection from cognitive decline through inhibition of the reduction in both gray matter volume and attentional performance associated with normal aging.

There is now evidence that meditation can also lead to changes in white matter, which is responsible for communication among the various regions of the brain. About this issue, Yi-Yuan Tang of Dalian University of Technology in China, University of Oregon (UO) psychologist Michael Posner, and their colleagues have used a type of MRI technique called diffusion tensor imaging (or DTI) to examine the impact of meditation on white-matter connectivity between cerebral structures.

In this study, forty-five UO students were divided in two groups. In one group, twenty-two students received eleven hours of integrative body–mind training (IBMT), an approach based on traditional Chinese medicine that involves mindfulness, mental imagery, and body relaxation. In the other (control) group, twenty-three students received the same amount of relaxation training involving the relaxing of muscle groups over the face, head, shoulders, arms, legs, chest, back, and abdomen. The participants were scanned before and after training. A comparison of the scans taken of the participants' brains before and after the training showed that those in the IBMT group had increased connections in the area of the anterior cingulate cortex, a region that plays an important role in attention and the regulation of emotions. White-matter changes were not observed in the control group.[34]

▐▐▐▐▐

So far, most brain imaging studies of meditation have not enabled researchers to determine whether differences between experienced meditators and novices existed before the studies. One way to address this vital issue is to investigate novice meditators and matched nonmeditator controls and follow them prospectively through time. Such a methodological strategy has been used by the scientists leading the Shamatha Project, the most extensive investigation to date regarding the benefits to mental and physical health of intensive meditation practice. This project, which is headed by Clifford Saron, a neuroscientist at the University of California–Davis, investigates the psychological and physiological processes underlying the long-term beneficial effects of meditation. Funded by the Fetzer Institute and the Hershey Family Foundation, the Shamatha Project involves a team of several psychologists and neuroscientists from universities across the United States and Europe.

In this project, sixty healthy people with varying degrees of prior meditation experience were randomly assigned to an intensive three-month meditation retreat or a control group. The control participants later followed a similar three-month retreat. Perceptual, cognitive, and emotional tasks, as well as questionnaires and physiological tests, were used to assess participants before, during, and after their retreats. During the retreat, participants received instruction in meditative techniques aimed at refining attention, and developing compassion and kindness toward others. Participants practiced alone about six hours a day over the three-month period. Meditation instructions were given by B. Alan Wallace, who combines fourteen years of training as a Tibetan Buddhist monk with degrees in physics, the philosophy of science, and religious studies. Founder and president of the Santa Barbara Institute for Consciousness Studies,

Wallace is currently one of the main proponents of the integration of Buddhist contemplative practices and Western science to move forward the study of the mind and consciousness.

The initial results indicate that intensive meditative training enhances attention, improves well-being, and promotes more empathic emotional response to the suffering of others. These positive changes endured at least five months after the retreat. The data related to the impact of the meditative techniques on brain activity are currently being analyzed and should be published soon.[35]

▥

Since the beginning of neuroscience in the nineteenth century, neuroscientists thought that we are "stuck" with the brain we are born with because they conceived this part of the central nervous system as a stable, hardwired machine. With the discovery of neuroplasticity, however, it became clear that this belief was deeply flawed. Indeed, researchers have found out that the adult brain is highly malleable.

Today, this evidence is impossible to ignore. Research has shown that we can intentionally train our minds, through meditative practices, to bolster the activity of regions and circuits of our brains involved not only in attention and concentration, but in empathy, compassion, and emotional well-being. Such mental exercises can even modify the physical structure of the brain. Changes in thoughts, beliefs, and emotions, made in the context of psychotherapy, also have the power to transform the brain, as shown by neuroimaging studies. Additionally, there is now some evidence that mental training can slow down the cognitive decline and reduction in gray matter volume typically seen in normal aging. These cutting-edge findings are great news. They invite us all to unleash the full potential of the mind, the immense power that sits within us.

Can we use the power of our minds to cure disease? In one study of this question, Dr. Carl Simonton and his colleagues taught patients to visualize their bodies in perfect working order and mentally imagine white blood cells as sharks devouring and eliminating the cancer cells, imagined as shark bait.[36] Results revealed increased life expectancy, better pain management, more positive attitude and self-images, and reduced tumor size and incidence for those patients who used the visualization technique. Similar results were found in other studies. In the next chapter we look at these and other intriguing interactions among the mind, the brain, and the body, and the very real curative potential that exists within us.

4

Surfing the Psychosomatic Network

The Intimate Connection of Mind and Body

All of your body is in your mind, but not
all of your mind is in your body.

—HUMANITARIAN, AUTHOR, AND
YOGI JACK SCHWARZ[1]

In 1973, a group of Indian researchers decided to test the extraordinary claim of some yogis that they could voluntarily stop their hearts—and survive. In this fascinating experiment, Yogi Satyamurti, a small man about sixty years old, was buried for eight days in a small underground pit dug into the lawn of a medical institute. He could not move, but he was connected to an electrocardiogram (EKG) device to record his heart's electrical activity. Beforehand, he had told the researchers that he would fall into a deep trance, from which he planned to awaken in seven days—a full eight days after he was buried alive.

Yogi Satyamurti climbed into the pit, which was then sealed with bricks and cement mortar. Almost immediately, the EKG showed a rapid heartbeat, called tachycardia, that progressed until it reached 250 beats per minute—far above the normal resting heart rate of 60 to 100 beats per minute. This tachycardia continued for an astounding twenty-nine hours.

Then what researchers had feared suddenly happened. A straight line—indicating that his heart had stopped—appeared on the EKG tracing. The researchers wanted to abort the experiment. Clearly, the yogi was dead or dying. But the yogi's attendants insisted that they continue for the full eight days.

For five more days, the yogi remained in the pit and the EKG continued to show a flat line. Then, half an hour before the experiment was scheduled to end, the needle began to move and the rapid heart rate appeared again. At the appointed time, they unsealed the pit and brought the yogi out—ten pounds lighter but otherwise alive and well.

His rapid heartbeat persisted for another two hours, and then returned to normal. The EKG device was checked to eliminate any malfunctioning, but it appeared to work flawlessly. The researchers were unable to account for this remarkable finding, but they admitted that they were not ready to accept that the yogi had deliberately stopped his heart for five days and survived.[2]

"The more optimistic amongst us considered this feat to be a marvellous extension of the 'hypometabolic wakeful state of yogic meditation' . . . and the conditioned learning of autonomic responses in rats reported by DiCara," said one of the researchers in a letter to the *Journal of the American Heart Association* in 1973. "The sceptics, however, were inclined to take the whole thing as some cleverly disguised trick. But, for the present, we only want to put this interesting experiment on record just as an intriguing and inclusive attempt of a Yogi to demonstrate a voluntary control over his heart beat."[3] His statement demonstrates the difficulty Western science and medicine have traditionally had in coming to terms with the close relationship between mind and body, and the possibility that we can indeed use our minds to influence our health. Yet there is good evidence that such things are indeed possible.

IIII

In 1964, a few years before the yogi went into the pit, Norman Cousins—political journalist, peace activist, and editor-in-chief of the *Saturday Review* for more than thirty-five years—found himself in a great deal of pain. He was diagnosed as having ankylosing spondylitis, a chronic form of inflammatory arthritis that causes the breakdown of collagen[4] in the joints between the vertebrae of the spine. "In a sense," he has written, "I was coming unstuck, I had considerable difficulty in moving my limbs and even turning over in bed. . . . At the low point of my illness, my jaws were almost locked."[5] Cousins was told that he had only a few months to live and that he needed to get his affairs in order. Instead, he embarked on an unusual regimen of his own devising.

While hospitalized, Cousins read about the theory that stress and negative emotions are harmful to the body. He reasoned that if negative emotions were damaging, then positive emotions should improve health. Shocked by hospital conditions and persuaded that the authoritarian medical culture was not going to be good for his health status, he decided to fire his treating physician and find a doctor who would agree to work with him as a collaborator rather than as the physician "in charge."

Cousins checked out of the hospital and checked into a Manhattan hotel. He began taking extremely high doses of vitamin C, strongly convinced of the efficacy of this treatment although his new physician did not endorse this approach. Decades before videos were commonly available, Cousins managed to obtain a movie projector and a stack of comedies, including Marx Brothers movies. Despite his pain he spent a lot of time watching these films and laughing. He quickly realized that ten minutes of genuine belly laughter had an anaesthetic effect and would give him at least two hours of pain-free sleep.

Over the next few months, Cousins slowly recovered the use of his limbs and his health gradually improved. Later, he returned to work full-time at the *Saturday Review*. He recounted his journey in a book titled *Anatomy of an Illness as Perceived by the Patient* "[6] In this book Cousins concluded that the "will to live" and positive emotions could have a huge impact on health, and even contribute to healing. He became popularly known as "the man who laughed himself to health," and his book brought widespread attention to the idea that mind and body are not separate but are intimately connected.

If Cousins were the only person to test this theory, his experience would be extraordinary. But he is by no means unique. Take, for instance, the case of Jordan Fieldman.

In the 1980s, this Harvard medical student—already suffering from ulcerative colitis, a painful form of inflammatory bowel disease—was diagnosed with an aggressive brain tumor and had surgery to remove it. Unfortunately, he woke from surgery blind, and he was told that he would remain sightless for the few months he had left to live. But a week later, his sight returned. When he went to the medical library to research his disease, every book he consulted asserted that recurrence is invariable and death occurs within a year. Fieldman, who had already experienced the unexpected and unpredicted return of his sight, was incensed. "How dare they say invariable?" he thought to himself. At that moment he simply decided not to die of his brain tumor. With that decision, his ulcerative colitis also regressed.

Combining traditional medical therapy with alternative health approaches, Fieldman did indeed live—just as he had decided. He graduated from medical school in 1987 and specialized in preventive and internal medicine, focusing on a holistic approach. In 1997, Fieldman told the *Boston Globe,* "When Western medicine gives you zero percent survival, you start exploring other options. It would have been easy for me to obey the odds and do what it says in the textbooks. But I had faith I could overcome it."[7]

Bernie Siegel, a retired New Haven surgeon, has written abundantly about empowering patients and teaching survival behavior. He believes that the power of healing stems from will, self-love, and hope. During his long career as a surgeon, Siegel had been personally involved with several patients, including Fieldman, who survived against all odds and experienced remarkable recoveries from supposedly incurable diseases. Of the mind–body connection, he writes, "Our bodies love us but if we do not love them and our lives then our body tries to get us out of here as fast as it can."[8]

Psychosomatic medicine—founded on the belief that the mind can cause bodily symptoms—emerged during the beginning of the twentieth century in reaction against the mechanistic view that had taken over science and medicine.[9] The Hungarian American psychoanalyst and physician Franz Alexander, one of the founders of this field, claimed that psychological factors are crucially involved in the production of disease. Many physicians and scientists could not see a plausible mechanism to explain the link between mind and body, and dismissed this idea as nonsense.[10] The idea persisted; but it would be decades before research became more than speculative.

In the 1960s, George Solomon, a psychiatrist working at Stanford University, began conducting research on the effects of psychological factors, such as personality characteristics and emotional states, on the onset and course of rheumatoid arthritis. In 1964 he published a seminal article titled "Emotions, Immunity, and Disease: A Speculative Theoretical Integration,"[11] in which he argued that the immune system could be influenced by mental events associated with brain activity. To describe this, Solomon coined the term *psychoimmunology*.

Our immune systems are composed of the lymph nodes, the spleen, the bone marrow, the thymus, and various types of white blood cells. Some of these cells reside in certain tissues of the body, such as the skin, while others circulate throughout the body. The main job of the immune system is to recognize, bar, and destroy foreign agents of diseases that threaten our health. It also disposes of abnormal cells and repairs damage. To do this, the immune system must determine and discriminate between something that needs to be fixed and something that needs to be destroyed. Autoimmune diseases, such as the anklyosing spondylitis that Norman Cousins had, occur when something goes wrong and the autoimmune response mistakenly identifies its own cells as intruders and tries to destroy them.

By the early 1970s, clinicians and researchers realized that if stressful situations and stressed-out personalities were conducive to a variety of illnesses, including heart disease and cancer, it was quite reasonable that reducing stress might help prevent these illnesses and contribute to more efficient treatments.[12] In 1981 neuroscientist David Felten[13] and his colleagues found a direct connection between nerve fibers of the sympathetic nervous system and cells of the immune system in the spleen, lymph nodes, thymus, and bone marrow. These researchers provided the first indication of how the brain and the immune systems can interact. This discovery established the field of psychoneuroimmunology (PNI), the study of the interactions between mental processes and the nervous and immune systems.

A few years later, neuropharmacologist Candace Pert and her co-workers discovered that neuropeptide receptors are present in the immune system. Neuropeptides—composed of short chains of amino acids—are small molecules that are used by neurons to communicate with each other. These molecules are implicated in various functions, including emotions, motivation, learning, and memory as well as food intake. Pert's discovery suggested a viable mechanism through which emotions can influence the immune system.

Since the crucial findings of Felten and Pert, PNI researchers have demonstrated that there are, in fact, a myriad of connections between the brain and the immune system. Studies conducted during the past three decades have shown that chemicals produced by immune cells signal the brain, and the brain sends chemical signals to the immune system.[14] Other studies have confirmed that our thoughts and feelings do affect our health and well-being. These studies indicate that the causes, development, and outcomes of an illness are determined by the interaction of psychological and social factors with biochemical changes that affect the immune system, the endocrine system, and the cardiovascular system.[15]

Here's how we think it works: the mind, the nervous system, the immune system, and the endocrine system form a psychosomatic network. They are continuously communicating via chemical messengers, such as neuropeptides, that can be thought of as "information substances." These chemical messengers allow subjective mental events—falling in love, losing a job, arguing with a friend—to influence, usually unconsciously, these physiological systems. But is it possible for the mind to deliberately affect this vast multidirectional network?[16]

Under ideal conditions, the mind and the nervous, immune, and endocrine systems interact harmoniously. The result is homeostasis, an optimal state of balance that promotes health and combats disease. Many factors influence the interaction between mind and these systems, such as lifestyle, environment, personality, and heredity. Illness may appear when there is a serious disruption of homeostasis.[17]

Increasingly, we are seeing evidence that the mind also affects gene expression—the process of how a gene works within a cell. Each cell in the human body contains several thousand genes, but not all of them are active at the same time. Within any given cell, some genes will be expressed (or "on") while other genes will not be expressed (they are said to be "off"). Genes that are turned "on" may somehow alter, for instance, our body's response to stress.

Psychologist Jeffery Dusek and his colleagues at Harvard Medical School recently conducted the first study about how the mind can affect gene expression. They compared gene-expression patterns in nineteen long-term relaxation practitioners, nineteen healthy controls, and twenty newcomers who underwent eight weeks of relaxation training. Dusek and his co-workers found that more than 2,200 genes were activated differently in the longtime practitioners relative to the controls. They found that a gene that is turned on or off by stress is turned the *other* way during relaxation. As the authors state in their conclusions, these "constitutive changes in gene expression . . . may relate to long-term physiological effects."[18] These findings have significant implications for the power of mind and body working in tandem to affect our health on a deep level.

▌▌▌▌

A considerable portion of the human molecular and cellular machinery is employed to maintain homeostasis—the body's ability to regulate itself in all sorts of situations. When we imagine consciously having to direct our bodies to perform all of the minute adjustments and functions that get us through twenty-four hours, we can see just how impressive and important this task is. When this state of equilibrium is perturbed or threatened, the brain's stress response system quickly springs into action. When we become stressed out, the brain releases stress hormones and chemical messengers (such as adrenaline and noradrenaline), and the adrenal glands secrete cortisol.

These biochemical reactions affect the capacity of immune cells to fight infectious agents.[19]

Although some stress is inevitable in daily living, and a degree of stress is necessary for survival, too much stress can have a detrimental effect on immunity. Acute or chronic psychological stress can increase the severity and duration of infectious diseases, prolong wound healing, and stimulate the reactivation of latent viruses. For instance, it has been shown that medical students receiving hepatitis B vaccine during the very stressful final exam period do not develop complete protection against this serious infectious disease. Moreover, negative immune changes have been documented for several months in victims of natural disasters such as hurricanes. Imprisonment in a prisoner of war camp and unemployment can also lead to negative immune alterations.[20]

Events and situations that we perceive as uncontrollable can lead to important disruption of the immune and endocrine systems.[21] For example, during marital discord, immune responses are weakened; and the levels of stress hormones are elevated in the spouse who experiences the greatest amount of stress and feelings of helplessness.[22] Furthermore, men and women who provide long-term care for a spouse or parent with Alzheimer's disease commonly show high levels of stress hormones as well as modifications in vaccine response and wound healing.[23]

A number of studies have shown that chronic negative emotional states may even contribute to death. For instance, a study of 2,400 patients in Finland has shown that the feeling of "hopelessness" was significantly associated with higher rates of mortality from cancer and cardiovascular disease.[24] Abnormality in immune function and an increased risk of death have also been reported in bereaved spouses during the first years after bereavement.[25]

Fortunately, positive emotions and determination can substantially improve health. One illustration of the impact of the "will to live" is that some people appear to have postponed their deaths until after some meaningful event, such as the arrival of a significant day or a loved one.[26] Two famous examples are that of Thomas Jefferson and John Adams. These two influential founders of the United States died on the same day—July 4, 1826—exactly fifty years after the Declaration of Independence was adopted.[27]

Researcher Sheldon Cohen and his colleagues at Carnegie Mellon University have recently performed studies in which volunteers were administered standard doses of infectious organisms such as rhinovirus (common cold) and influenza virus. The responses to the viruses were analyzed in relation to the emotional state of the participants. In these studies, volunteers were monitored in quarantine. Results revealed that participants whose affect remained very positive over several days had reduced risk of developing an infection.[28]

Is there any evidence that an individual's attitude and emotional state can positively influence the course of diseases more dangerous than the common cold and influenza? To tackle this important health issue, a group of researchers studied women with breast cancer, all of whom had undergone a simple mastectomy. Three months after the surgery, the researchers interviewed each woman to find out what having cancer implied to her, and to ask what she thought about the threat of cancer. These researchers discovered that at the five-, ten-, and fifteen-year follow-ups, the best single predictor of recurrence of cancer or death was the mental attitude of each woman three months after mastectomy. Women who showed fighting spirit had a 50 percent chance of surviving fifteen years in good health; whereas women who accepted stoically, or felt anxiety or helplessness, had about a 15 percent chance of surviving fifteen years.[29]

Emotional support also plays a pivotal role in health. Individuals with high levels of support have stronger natural killer (NK) cell responses[30] than do individuals with lower levels of support; and high-quality emotional support from a spouse is associated with better NK cell activity in breast cancer patients.[31]

The effect of emotional support on people who have already been diagnosed with cancer can be dramatic. Studies indicate that people with various types of cancer who have more friends and relatives whom they see have a better quality of life and tend to live longer. One of these studies revealed that women with breast cancer who had no confidants had a seven-year survival rate of 56 percent, whereas those who had two or more confidants had a survival rate of 76 percent.[32]

Cancer patients sometimes use visualization techniques as part of their treatment. Mind–body techniques such as mental imagery, relaxation, and deep breathing are relatively new to the West, but familiar to much of the world. Traditionally, they have played an important role in African, Indian, Chinese, and Native American medicine, where they are predicated on the view that our thoughts, emotions, and beliefs can affect several aspects of our bodies' functioning. It is only in the past thirty years that Western medicine has begun to recognize the importance of such techniques.[33]

Mental imagery—the ability to form mental images of objects or events when the relevant objects or events are not present to the senses—is one of the oldest healing techniques. It includes all senses—sight, hearing, touch, taste, and smell—and can also be used to simulate a given action. Brain imaging studies have shown that these various forms of mental imagery activate specific areas of our brains as effectively as if the people scanned were actually seeing, hearing, touching, tasting, smelling, or moving.

In the 1990s, Howard Hall, a research psychologist at Case Western University in Ohio, showed that healthy volunteers could use mental imagery to positively influence the functioning of their immune systems, in particular the activity of neutrophils, the most abundant type of white blood cells.[34] This important finding demonstrated that even if a great deal of the activity of the psychosomatic network takes place at the unconscious level, it is possible for the mind to consciously influence its functioning.

Radiation oncologist Carl Simonton and psychologist Stephanie Matthews Simonton were the first to propose the use of relaxation and mental imagery to enhance the capacity of the immune system to defend against cancer. They encouraged people with cancer to visualize images of immune cells aggressively conquering or destroying cancer cells. Since the pioneering work of the Simontons, however, research has shown that visualizing a battle in which white blood cells destroy cancer cells does not serve or suit everyone. The types of images that are most efficient vary from individual to individual. For example, some people feel empowered by warlike images, while others prefer relaxing images such as cancer cells packing up and leaving the body. Moreover, some people prefer images that are symbolic, such as a big broom sweeping up cancer cells, whereas other people prefer images of white blood cells and cancer cells that are anatomically accurate.[35]

Several studies that investigated the effect of mind–body techniques on people who have cancer involved a combination of mental imagery and relaxation. A vast majority of these studies indicate that mental imagery combined with relaxation increases the production and activity of immune cells, reduces nausea and vomiting associated with chemotherapy and the distress of radiation therapy, facilitates recovery from cancer surgery, decreases anxiety and depression, and enhances quality of life.[36]

Mindfulness meditation is another mind–body technique that can help individuals suffering from cancer to cope with the physical and emotional symptoms associated with this disease. Research shows that people with various kinds of cancer who practice mindfulness on a daily basis report less anxiety, depression, anger, and bodily manifestations of stress.[37]

||||

Dissociative identity disorder (DID)—previously known as multiple personality disorder—is a condition in which a person appears to possess and manifest distinct, split personalities: a *host personality* or *host,* and one or more *alter egos* or *alters.* These distinct personalities alternate in control over the person's thoughts, memories, feelings, conscious experience, actions, and sense of identity. Usually, the person with DID is not aware of the other personality states and does not have memories of the times when other alters are in control.[38]

The number of alters in persons with DID varies widely. For example, the woman known as "Eve," Chris Costner Sizemore—whose story was told in the 1957 film *The Three Faces of Eve—*

developed a total of twenty-two different personalities over twenty years.

DID is relatively rare: its prevalence is about 1 percent in the general population. Nearly all of the individuals who develop this condition have experienced repeated trauma—overt physical, emotional or sexual abuse, or neglect—during their childhoods. Psychiatrists believe that dissociation is a coping mechanism; the person dissociates herself from events that are too traumatic, frightening, or painful to be assimilated within her conscious self.

Intriguingly, alters have their own ages, gender, name, race, or sexual orientation distinct from the host. They also have distinct postures, gestures, ways of talking, and temperaments. The transition between alters, called the "switch," can take seconds to minutes to days.

In 2008, Herschel Walker, a former Heisman Trophy winner and NFL football star, released a book titled *Breaking Free,* in which he revealed that he suffers from DID. Walker's life derailed not long after his football career ended. "I didn't really learn about this [DID] until about ten years ago," Walker told CNN reporter Miriam Falco. "My life was out of control. I was not happy, I was very sad, I was angry and I didn't understand why."[39]

In his book Walker talks about a dozen alters, including the Hero, the Warrior, the Enforcer, the Daredevil, and the Consoler. Some of these alters contributed in generating good things: they helped Walker graduate at the top of his class and become one of the greatest college football players ever. But other alters caused wild and violent behaviors. On a number of occasions, Walker put himself in danger playing Russian roulette. At another time, the very late delivery of a car enraged him (or, more precisely, one of his alters) so much that he thought about killing someone. Ultimately, these violent alters led to the breakup of his nineteen-year marriage to Cindy Grossman. "I lost the person that was like everything to me," Walker said. "I lost my wife and that's totally, totally devastating to me." [40]

Grossman said that when the "bad" alters emerged, her husband looked different. His eyes, she said, would turn "evil—I remember just getting chill bumps when he looked at me." [41]

A handful of times, Walker held a gun to his wife's head and threatened her with knives. Once, Grossman was in bed and could not see well because she wasn't wearing her contact lenses. Abruptly, she said, Walker threatened her with a razor. "He had it to my throat and kept saying he was going to kill me . . . think he choked me. I think I passed out." When she regained consciousness, "There was someone else there [saying] 'Cindy, Cindy, Cindy. Wake up, wake up!' " Walker said that he has no memory of these assaults.

Walker believes his mental condition is related to a trauma that occurred while he was in elementary school. As a fat child with a severe stuttering problem, he was beaten up a lot by other kids and couldn't fight back. [42]

One fascinating aspect of DID is the fact that the distinct alters often exhibit markedly different symptoms or physiological features. A few preliminary studies have been conducted to investigate this intriguing issue. For example, some of these studies have shown that one alter can experience anaesthesia or pain relief, but the others cannot. There have been reports of one alter being deaf or suffering from auditory hallucinations, while the others hear normally. Changes in handedness or handwriting have also been reported across alters.

Sometimes individuals with DID display allergies in some alters but not in others. For instance, one of Eve's alters ("Eve Black") showed an allergic reaction when she was wearing nylon stockings. Strikingly, Eve's original personality ("Eve White") did not. In another study one alter was allergic to cats, while another was not. Additionally, alter egos can respond differently to medications. For instance, a woman who had diabetes needed different amounts of insulin depending on which alter was in charge.

Changes in vision have also been commonly described. In one study, major optical differences—evidenced by measures of visual acuity, pupil size, corneal curvature, and so on—were found between alter egos. In another investigation, color-blindness was reported in one alter only. Research involving the measurement of EEG activity elicited by the presentation of pictures suggests that different alters can be associated by distinct brain responses. In other respects, a group of Dutch neuroscientists have used PET to demonstrate that different senses of self are associated with different patterns of regional cerebral blood flow in various regions of the brain.[43]

Recurrent traumatic wounds are seen in certain individuals with DID. In one case, a woman whose mother had burned her with cigarettes over and over again displayed red marks on her skin when one alter surfaced. In another case, a woman addicted to heroin in one personality, but not in the others, exhibited needle track marks when she switched to that alter.[44]

Clearly, switches between different alters can be accompanied by distinct physiological changes. Researchers are still puzzled by this tantalizing phenomenon. The bodily alterations associated with switches support the view that changes in mental states and events are accompanied by modifications in physiological activity. In other words, *change the mind and you change the brain and the body.* It is noteworthy that these modifications disappear following successful psychotherapeutic integration of the various alters.

||||

Yogi Satyamurti, whose story was told at the beginning of this chapter, appeared to have stopped his heart for several days and then started it again. Other yogis have demonstrated that it is possible to voluntarily influence the temperature of the body through the practice of *tummo* meditation. Harvard researcher Herbert Benson found that practitioners of this form of meditation, who had learned to warm themselves while meditating in cold Himalayan mountains, could deliberately raise the temperature of their fingers and toes, by amounts ranging from 31.5°Celsius (C) up to 8.3°C, or 89°Fahrenheit (F) to 47°F. Benson proposed that these yogis were probably able to deliberately influence the widening of blood vessels (vasodilatation) in their toes and fingers.[45]

Impressive voluntary control over bodily processes that usually are not supposed to be influenced voluntarily has also been seen in Westerners, and not in conjunction with yogic practices. Helen Flanders Dunbar, a psychoanalyst who specialized in psychosomatic medicine, described the cases of two Americans who could apparently produce urticaria, or wheals (raised red areas) on the skin, at will. One was able to generate an urticarial wheal at any selected spot on his forehead simply by "thinking about it." The other was a physician who was able to produce urticaria on his trunk and arms.[46]

In the 1970s, physicist Elmer Green and his wife, Alyce, a psychologist, brought Dutch yoga practitioner Jack Schwarz into their laboratory at the Menninger Foundation in Topeka, Kansas. Schwarz told the Greens that by voluntarily controlling his brain waves, he could control his blood pressure, heart rate, and the pain of physical trauma. Controlling his brain activity, he explained, involved moving his attention toward his "higher self," the omnipotent and infinite intelligence within him.

Schwarz demonstrated the power of the mind over the body by putting a long, unwashed sail-maker's needle through his biceps with no resultant pain, bleeding, or infection.[47] At the same time, his brain's electrical activity, which was measured with an EEG device, dropped suddenly into the alpha band.

||||

Today, the concept of a psychosomatic network requires mind and body to constitute an indivisible unity. This holistic conception has helped greatly to diminish the resistance of physicians and scientists to the idea that psychological factors can play an important role in health. For many contemporary PNI researchers, the psychological component of the psychosomatic network can simply be reduced to the brain. Oakley Ray, a professor of psychology and psychiatry at Vanderbilt University, writes that "our thoughts, our feelings, our beliefs, our hopes, are nothing more than chemical and electrical activity in the nerve cells of our brain." [48]

The words "nothing more" are always a signpost to the reductionist view of life—in this case, the belief that the mind is identical with brain activity. I agree wholeheartedly with Emily Williams Kelly, a research psychologist at the University of Virginia, that this reductionist view avoids the problem of how subjective, nonmaterial mental events such as thoughts, emotions, and mental images can lead to very specific changes in the objective, physical body.[49] This simplistic perspective fails miserably to explain how, for example, people can intentionally stop their heart or mentally produce wheals on their skin.

This transduction process—the conversion of a signal from one system to another, in this case, from the mind to the body—still remains very mysterious. Candace Pert has hypothesized that information is the missing element that allows us to transcend the Cartesian mind–body split since it touches both mind and body. Indeed, information is processed both by the mind and the body, and it is not dependent on time and space, as matter and energy.[50] It is also possible that a larger Individuality or Self, of which our ordinary waking mind is usually not aware, is involved in the transduction of mental events into specific bodily changes. This idea was proposed a century ago by the brilliant Frederic Myers.[51]

Hypnosis is now a fairly common tool in health management—it is used, for example, by people who wish to quit smoking or get past a fear of dentists so that they can take better care of their teeth. Astonishingly, despite our familiarity with it, science still has no definitive idea of what it is or how it works. And, as the next chapter demonstrates, the ability of hypnosis to harness the mind's power is being explored and utilized in some surprising ways.

5

The Mind Force Within

Hypnosis

People think hypnosis is about giving up control. But
it's actually giving control back to the patients.

—PSYCHIATRIST DAVID SPIEGEL[1]

When John was born, his mother could see that there was something wrong with his skin. As he grew up, his whole body, apart
from his face, neck, and chest, became covered by a thick, in-
elastic, black substance that bore no resemblance to normal skin.
This black substance was as hard as teeth or fingernails, and
John was troubled by painful cracking of the skin. He was a
horrible sight, and at school John was treated like an outcast.
Doctors told John's parents that he was suffering from *ichthyosiform erythrodermia,* a harrowing condition known as "fish skin
disease." Unfortunately, no cure was known and John's chances
of leading a normal life appeared negligible.

In May 1950, at age fifteen, John could barely move without causing painful fissures in his "black armour plating." He was admitted to the Queen Victoria Hospital in East Grinstead in Sussex, England, a hospital with an international reputation for its plastic surgery department. The famous plastic surgeon Sir Archibald McIndoe and his team would try to help.

McIndoe and his colleagues began with the palms of John's hands, which were now enveloped in "a rigid horny casing" that had cracked and become infected. The surgeons scraped the black substance off both his horny palms and transplanted some skin from his chest. But the operation was soon shown to have failed: a few weeks later the grafted skin had thickened and turned black. When a second attempt also proved unsuccessful, McIndoe and his associates concluded that they could not do anything more for John.

But anaesthesiologist Albert Mason had an innovative idea: he proposed to try hypnosis on John. Mason had previously cured warts through hypnosis and was firmly convinced that hypnotic suggestions could significantly improve the boy's condition.

Mason began by talking John into a hypnotic trance state. Then he repeatedly told the boy, "Your left arm will clear." Strikingly, five days later, "the horny layer softened, became friable, and fell off. The skin underneath became pink and soft within a few days. . . . At the end of ten days the arm was completely clear from the shoulder to the wrist."[2] Mason continued his treatment, beginning with the right arm and moving on to the trunk and the legs. The improvement was startling: it ranged from 50 percent on the legs and feet to 95 percent on the right arm.

Justifiably proud, Albert Mason took John along to show him to Archibald McIndoe. Flabbergasted, McIndoe told Mason that

the outcome of his hypnotic treatment made no sense medi-
cally—John's skin had no oil-forming glands that would allow
its outer layers to peel off and refurbish themselves. Yet McIndoe
was deeply impressed and proposed to Mason to present John at
the next Royal Society of Medicine (RSM) conference. Mason
agreed. The doctors who attended John's presentation at the
RSM were similarly baffled. Some said that John's case challenged the traditional medical concepts about the relation between mind and body.

This story has a happy ending. One year after the end of the
treatment, John had become a normal, happy young man and
had found a job as an electrician's assistant.[3] A few years later, the
case was re-presented to the RSM. Mason reported, "Not only
has there been no relapse, his skin has continued to improve. . . .
without further treatment of any sort, hypnotic of otherwise." [4]

John's recovery is vivid and heartening. It illustrates particularly well how strong the mind can sometimes be and to what
extent hypnosis can allow us to tap into the power within. But
what is hypnosis? And how does it work?

More than half a century after John's early miraculous recovery, there is still no consensus regarding the hypnosis or how
it works. We can say that hypnosis is a trance state (that is, an
altered state of consciousness) that is accompanied by changes in
perception, memory, emotions, and action—changes that might
seem familiar. Indeed, spontaneous mild trance states constitute
a normal activity of the mind. These states occur routinely—
for example, when we are daydreaming, reading an interesting
book, or watching a captivating movie.

Most experts agree that hypnosis is a cooperative social inter-action in which one person, the subject, becomes highly focused and receptive to verbal suggestions given by another person, the hypnotist. In the induced hypnotic state, however, the mind is guided by the hypnotist, who makes specific suggestions to help the subject change a medical condition, a perception, or a be-havior. To guide the subject into a hypnotic state, the hypnotist generally asks the subject to close her eyes, breathe in a certain way (designed to calm and regulate the breath), and visualize a relaxing scene (for example, "Picture yourself walking along a long, smooth, white beach . . . hear the waves breaking on the shore . . ."). In this way the subject becomes relaxed and open to the hypnotist's suggestions.

But hypnosis does not necessarily have to involve another person. Hypnotic suggestions can also be self-administered: a hypnotic state that is self-created is called *autohypnosis* (or self-hypnosis). Many experts believe that if you accept the sugges-tions made by a hypnotist, you are actually hypnotizing yourself. In other words, all hypnosis would be essentially self-hypnosis, even when it is assisted by a hypnotist.

Because hypnosis is a continuum rather than an all-or-nothing phenomenon, most people can be hypnotized to some degree.

Many people, however, insist that they *can't* be hypnotized. A number of reasons exist for this attitude, including the belief that hypnosis requires the subject to be weak minded, naïve, or unintelligent. But this belief is mistaken. Indeed, research shows that the best candidates for hypnosis are intelligent and focused. Researchers have found that a propensity to become absorbed in fantasy or imagery—such as in books or movies—and an aptitude for blocking out the surrounding world are the best predictors of the capacity to be hypnotized.[5]

Some people also believe, erroneously, that a person under hypnosis can be forced to do things against her or his will. This attitude has not been helped by nightclub acts in which stage hypnotists seem to compel audience volunteers to do embarrassing things—cluck like a chicken, cry like a baby, or act foolishly in some other way. In fact, during the hypnotic state, hypnotized subjects do not lose complete control over their behavior and do not do anything that they would not be inclined to do otherwise.

||||

The modern era of hypnosis began with the Scottish surgeon James Braid in the early nineteenth century. Using eye-fixation techniques, Braid was able to induce trance states. He also discovered the importance of using suggestions and is credited with introducing the term *hypnotism*—the act of inducing hypnosis.[6]

In the mid-nineteenth century, anesthetics were not yet available, and terrible pain was associated with every surgical procedure. The English surgeon John Elliotson—a professor of medicine at University College, London, and his protégé, Scottish surgeon James Esdaile, a medical officer for the British East India Company—proposed that hypnosis could be used as a therapeutic tool to reduce surgical pain. To test the validity of their proposal, Elliotson and Esdaile performed hundreds of surgical interventions with the hypnotic trance as the sole anesthetic agent.

Some of the surgeries Esdaile carried out in Bengal between 1845 and 1851 included amputations of breasts, penises, limbs, and scrotal tumors (caused by filariasis, an infectious tropical disease transmitted by mosquitoes). Here is how Esdaile describes a scrotal tumor case involving Gooroochuan Shah, a forty-year-old shopkeeper:

He has got a "monster tumor," which prevents him from moving; its great weight, and his having used it for a writing-desk for many years, has pressed it into its present shape. His pulse is weak, and his feet oedematous, which will make it very hazardous to attempt its removal; but with such an appendage, life is literally a [burden]. He became insensible on the fourth day of mesmerizing [hypnosis], and was drawn with the mattress to the end of the bed (my usual mode of proceeding): two men then held up the tumor in a sheet, pulling it forward at the same time, and in the presence of Mr. Bennett, I removed it by a circular incision, expedition being his only safety. The rush of venous blood was great, but fortunately soon arrested; and after tying the last vessel, the mattress was again pulled back upon the bed with him upon it, and at this moment he awoke. The loss of blood had been so great that he immediately fell into a fainting state, and it took a good while to remove him. On recovering he said that he awoke while the mattress was being pulled back, and that nothing had disturbed him. The tumor weighed 80 pounds, and is probably the largest ever removed from the human body. I think it's extremely likely that if the circulation had been hurried by pain and struggling, or if the shock to the system had been increased by bodily or mental anguish, the man would have bled to death, or never have rallied from the effects of the operation. But the sudden loss of blood was all he had to contend against; and, though in so weak a condition, he has surmounted this, and gone on very well.[7]

Remarkably, these surgeries performed under hypnosis were conducted without pain and with very low morbidity rates: in fact, the death rate was reduced to only about 5 percent at a time when surgical mortality was about 40 percent. Esdaile attributed this positive outcome to the successful relief of pain with hypnosis.[8]

Not long after the pioneering work of Elliotson and Esdaile, chemical anesthetics such as ether and chloroform became popular and displaced the use of hypnosis for anesthesia in surgery. Still, many observers had taken notice of the potential usefulness of hypnosis.

In the early part of the twentieth century, hypnosis became an object of scientific investigation, and American researchers interested in this emerging field became more influential than their European counterparts. Research on hypnosis exploded in the 1960s. Today scientific research into hypnosis is still growing, and hypnosis is becoming widely used in various areas of clinical psychology and medicine. It is now available to patients at some of the best medical centers in the United States, including Mount Sinai Medical Center and Beth Israel Medical Center in New York City, the Cleveland Clinic in Ohio, and Stanford Hospital on the campus of Stanford University in Palo Alto, California.[9]

▐▐▐

Contemporary research shows that what goes on in our minds can greatly influence our unpleasant sensations of pain. In fact, the psychological aspect of pain—how much pain we expect to feel—seems to be as important as the physiological processes that give rise to it. No wonder, then, that hypnosis can be used to induce pain relief.

Hypnotically induced analgesia—pain relief—represents one of the most dramatic, clinical applications of hypnosis. Although chemical anesthetics largely replaced the use of hypnosis for anesthesia in surgery during the second part of the nineteenth century, hypnosis is making a something of a comeback in this area. An increasing number of people today prefer to undergo surgery consciously, under hypnosis, to avoid the wooziness and other unpleasant aftereffects associated with general anesthesia.

The case of Pippa Plaisted illustrates this trend. A few years ago, this forty-six-year-old woman underwent a breast cancer operation at Lister Hospital, London, without anesthesia or painkilling drugs, using only hypnosis to relieve pain. It seemed a daring thing to do, but Plaisted had good reason: she had previously had several operations, and after each one of them, the anesthetic drugs had left her feeling dizzy and nauseous for several months.

Just before Plaisted's surgery began, hypnotherapist Charles Montigue stood at the operating table and put her into a hypnotic trance, resting his thumb on Plaisted's forehead to monitor her hypnotic state. During surgery she had her eyes closed and she could hear the surgeon telling her, at each phase of the operation, what was coming next. Astoundingly, she reports that she did not feel any sensation at all during the operation. "The surgeon was cutting and sewing inside me, but I could not feel any sensation at all," Plaisted recalls. "After the operation, I felt tired, but there was no nausea or wooziness. I had a clear head and felt totally normal."[10]

Some surgical patients choose to combine hypnosis with a local anesthetic to avoid the debilitating knockout effect of general anesthesia. Marianne Marquis is one of these patients. In 2011 she was hypnotized ten minutes before surgery to remove her thyroid, a small gland located at the bottom of the neck that produces hormones involved in the control of the body's metabolism. The operation was performed at Cliniques Universitaires St. Luc in Brussels, Belgium. Marquis, fifty-three, was put into a hypnotic trance by her anaesthesiologist, Dr. Fabienne Roelants. While the surgeons were slicing her neck open, Marquis said later, she was "imagining squishing my toes in the sand and feeling water come up over them." She did not feel any pain during the surgery.

Marianne Marquis's case is not unique. Combining hypnosis with local anesthetics is becoming an increasingly popular option in Western Europe. This approach is used at Roelants's hospital for one-third of the surgeries carried out to remove thyroids and one-quarter of the breast cancer surgeries. Fabienne Roelants and her colleagues are planning to expand this approach to knee arthroscopies and plastic surgeries. A few plastic and facial surgeons in Germany also currently use hypnosis for pain control.

Doctors who support this approach have several reasons. Clearly, it cuts down on the need for anesthetic drugs. Proponents also argue that the recovery time of patients is faster, and their need for painkillers afterward is diminished. Other doctors, however, warn that hypnosis is not usable in surgical interventions implicating the heart or other internal organs because the pain would be unendurable. And for the surgeon, there's no possibility of error. George Lewith, a professor of health research at Southampton University, England, says bluntly, "If hypnosis doesn't work and you've got somebody's abdomen or chest open, then you're in big trouble."[11] Another limitation is the fact that while many people can by hypnotized to some degree, not everyone can be brought into a deep hypnotic state.[12] Nonetheless, research shows that about half of the population can obtain significant pain relief—a one-third reduction in felt pain—from hypnosis.[13]

This is welcome news for people suffering from serious burns, for whom pain is a major problem. Pain impairs wound healing and contributes to posttraumatic stress disorder; and severe pain can even lead to the development and maintenance of suicidal thoughts in burn patients after they are released from the hospital. Powerful analgesic drugs, such as morphine, are generally given to severely burned patients to help them cope with acute pain. But the doses required for controlling this extreme pain are often very large, augmenting the risk of negative side effects. Given this, some researchers have sought to evaluate whether hypnosis can benefit burn patients and diminish analgesic drug consumption.

In one study, patients with major burns admitted to the intensive care unit of the University Hospital in Lausanne, Switzerland, were asked if they would like to try hypnosis in addition to conventional treatment (including analgesic medications). Patients in the hypnosis treatment group reported lower pain scores and less anxiety than control patients, who received only the standard treatment. Patients in the hypnosis group also required fewer pain-relieving drugs afterward and displayed better wound healing than those in the control group.[14]

Pain is the most common symptom of bone cancer, in which cancerous tumors destroy normal bone tissue. In a carefully designed investigation, patients with advanced bone cancer were randomized to receive either weekly sessions of supportive attention or a hypnosis intervention. Patients assigned to the hypnosis intervention were given a minimum of four weekly sessions in which a hypnotic induction—including suggestions and instructions—was completed. They were also provided with an audiocassette tape recording of a hypnotic induction and were instructed in home practice of self-hypnosis. The patients in the intervention group showed an overall reduction in pain for all hypnosis sessions combined compared to those who received only supportive attention. Moreover, the effectiveness of self-hypnosis home practice was reported as 6.5 on a 0-to-10 scale.[15]

Hypnosis has also been shown to be efficacious with migraines, a common and very painful type of headache that is often debilitating. In one study, patients who had suffered with migraines for a minimum of one year were randomly assigned to receive either medication treatment or hypnotherapy. The hypnosis treatment consisted of six sessions at intervals of about two weeks. Hypnotic suggestions included experiencing less tension and less anxiety, and patients were also instructed to imagine the arteries in the head as becoming smaller and more comfortable. They were then asked to practice self-hypnosis on a daily basis.

Results showed that that the number of migraines per month was significantly lower in the hypnosis group. In the second six months of treatment, the hypnosis group averaged only 0.5 migraines per month compared with nearly three per month in the medication group. At one-year follow-up, ten hypnosis patients had experienced complete remission of migraines during the previous three months—that is, they had experienced no migraines at all—compared with only three patients in the medication treatment group.[16]

Chronic pain—pain for any reason that lasts for more than six months—is a common reason for seeking medical care. Unrelieved chronic pain can lead to substantial suffering, emotional distress, and physical limitations. Frustratingly, it often persists despite treatment with analgesics and physical modalities. Nearly 20 percent of American adults claim they suffer from chronic pain.

The findings of several studies indicate that hypnosis produces significant decreases in pain associated with various chronic pain problems, including fibromyalgia, arthritis, low-back problems, and temporomandibular disorder (TMJ)—a condition resulting from an inflammation of the temporomandibular joint, which connects the jawbone to the skull. These studies have generally found that hypnosis is more effective than nonhypnotic interventions such as attention, physical therapy, and education. Thanks to these findings, interest in hypnosis for pain management is now getting stronger.

||||

The pain associated with labor and delivery is described by many—if not most—women as the most extreme pain they have ever experienced. Although pharmacologic methods such as analgesic drugs and epidurals have been shown to be effective, these methods can be risky for both mother and child, and after a certain point in labor cannot be administered at all. It's not surprising, then, that many expectant mothers have reservations about drug-induced pain relief.[17]

Over the past twenty years, a number of studies have been conducted to evaluate the effectiveness of hypnosis in reducing labor pains. In one of these studies, pregnant women were divided into two groups. Half of the women were randomly assigned to receive hypnotic suggestions for relaxation and analgesia; the other half received breathing and relaxation exercises. The hypnosis treatment was audiotaped for the patients to listen to daily prior to delivery; women in the control group listened to a commercial pre-birth relaxation tape. The women who received hypnosis had shorter labor, used less pain medication, and had higher rates of spontaneous deliveries than did women in the control group. Women receiving hypnosis also reported less labor pain.[18]

Another investigation sought to assess whether hypnosis could diminish birth complications. Overall, 520 pregnant women who were still in either their first or second trimester were recruited. These women were randomly assigned to either a hypnosis treatment or a supportive psychotherapy. The hypnosis intervention emphasized suggestions for decreased anxiety and fear, increased feelings of relaxation, and confidence in the ability to cope with the pain of labor. The supportive psychotherapy intervention was based on discussions of issues related to pregnancy. Women in the hypnosis treatment group experienced less complicated deliveries and Cesarean sections than those in the supportive psychotherapy group. In addition, the women who received the hypnotic suggestions used analgesics and epidural anesthesia less frequently than those in the other group.[19]

Recently, the birth outcomes of women undergoing hypnotic preparation for labor and delivery pain were assessed in the obstetrics and gynecological service of the Women's and Children's Hospital in Adelaide, Australia. Seventy-seven women—some who were having their first baby (primigravid), and some who had given birth more than once (parous)—received up to four forty- to sixty-minute hypnosis training sessions after thirty-five weeks of pregnancy. During the first session, these women were taught how to use hypnosis. In subsequent sessions, they received suggestions for relaxation and analgesia. The participants in the hypnosis treatment group were compared with a control group of 3,249 women, matched for parity (that is, the number of live-born children and stillbirths a woman has delivered at more than twenty weeks of gestation) and gestational age. Women in the control group received standard medical care. Results revealed that primigravid women in the hypnosis treatment group used epidural anesthesia significantly less frequently and had a decreased need for drugs during labor compared with women in the control group.[20]

Taken together, these studies suggest that hypnotic preparation can be more effective than standard medical care in diminishing pain and analgesic use during labor and delivery—good news for women seeking to give birth naturally, without drugs and with as little pain as possible.

||||

Surgery is often a frightening prospect of unknown outcomes that are out of the patient's control. So it's not surprising that anxiety is a real problem for people anticipating surgery. More than half of presurgical patients fear the effects of anesthesia or the possibility that they will not wake up after surgery. Hypnosis has been used for decades to lessen the anxiety associated with surgery.

A recent study examined the effect of hypnosis on preoperative anxiety.[21] People undergoing outpatient surgical procedures were randomized into three groups: a hypnosis group, an attention-control group, and a "standard of care" control group. Participants in the hypnosis group were told during the suggestion phase that they would continue to feel relaxed and calm before the surgery. In the attention group, participants received attentive listening and support, without any specific hypnotic suggestions. Researchers assessed the participants' anxiety levels before and after intervention, and on entering the operating room. Participants in the hypnosis group reported a marked decrease of 56 percent in their anxiety levels on entrance to the operating room. By contrast, the attention group reported an increase of 10 percent in anxiety, and the control group reported an increase of 47 percent in their anxiety.

Breast cancer is the most common cancer among women today, and the second leading cause of cancer deaths in women (after lung cancer). Screening mammography allows early detection of this type of cancer, and ameliorates chances of successful treatment and survival. In the United States, more than 60 percent of women aged forty or older undergo mammography each year. Five to 10 percent of mammograms produce inconclusive or abnormal results that require follow-up biopsies in which doctors use a large hollow-core needle to get small breast tissue samples. This procedure is usually done under local anesthetic, which limits the use of intravenous drugs to decrease anxiety and pain.

Recently, a group of researchers compared the effect of self-hypnosis, structured empathic attention, or standard care on the anxiety levels experienced by 236 women undergoing breast biopsy with a large core needle. The women receiving only standard care reported a significant increase in anxiety, while the anxiety level did not change in the empathy group. Anxiety decreased significantly in the self-hypnosis group.[22]

There is also considerable evidence that hypnosis reduces anxiety associated with various dental procedures and oral surgery, such as multiple dental extractions or the excision of cancer cells on the cheek.

||||

Hypnosis has also been shown to have a positive effect on healing a variety of conditions. These range from external conditions, such as skin problems and burns, to internal conditions such as asthma.

John's dramatic experience with fish skin disease, described at the beginning of this chapter, is just one example of a broad array of skin conditions that can be substantially improved using hypnosis. Two common conditions, psoriasis and urticaria, seem especially susceptible to suggestion.

Stress and emotions are known to play a major role in the onset, exacerbation, and prolongation of psoriasis, an inflammatory skin disease characterized by the formation of reddish spots and patches, especially on the scalp, trunk, elbows, and knees. Cases of extensive severe psoriasis that had resisted twenty years of conventional treatment and that have demonstrated marked improvement (for example, 75 percent resolution) or full resolution when treated with hypnotherapy have been reported.

Moreover, a double-blind controlled trial using hypnosis as complementary therapy in psoriasis has shown significant improvement in highly hypnotizable individuals.[23]

Urticaria (also called hives) is a skin eruption characterized by temporary wheals of changing shapes and sizes. This eruption is caused by an allergic reaction, which is due to a hypersensitivity of the immune system. Allergic reactions—which occur in response to usually harmless environmental substances called allergens (such as pollen and house dust)—are associated with disproportionate activation of certain white blood cells, such as mast cells and basophils. In a study of hypnosis on fifteen patients with chronic urticaria of nearly eight years' average duration, six patients had resolved and another eight improved within fourteen months. Other investigations have demonstrated that hypnotized volunteers can significantly decrease the immediate wheal-and-flare reaction to an injected allergen in a skin test. In a particularly interesting study, participants under hypnosis who were asked to increase their reaction to one allergen while decreasing it to the other showed a significant difference in reactions between the two allergens.[24]

Hypnosis has also been utilized with success to foster healing of severe burns. For instance, one study has shown faster surgical incision wounds healing in a group of participants having received hypnotic suggestions compared to control participants.[25] In another investigation, hypnotic suggestions helped severely burned patients to heal quickly and without scarring. One of these patients was a man whose right leg had briefly been immersed up to the knee in melting aluminum.[26]

The impact of hypnosis on asthma, which can cause severe breathing problems, has also been evaluated on several occasions.

In one clinical trial involving patients with this respiratory disease, those who were assigned randomly to the hypnosis group had lower wheezing scores and less use of bronchodilators (devices used by many asthmatics to administer medication that eases breathing) than patients in the control group. The participants who responded best were more easily hypnotized, more compliant with practicing self-hypnosis techniques, and experienced a deeper level of trance.[27]

IIII

Structural changes in the body without surgery, using only hypnosis, arguably constitute the most spectacular manifestation of the power of hypnosis. In the 1970s, a few studies were performed on women who wanted to increase the size of their breasts. These women were put into a hypnotic trance and instructed to visualize themselves as they wanted to look. Amazingly, nearly all of the women succeeded in increasing their breast size: across the studies, the average increase in the circumference of their breasts ranged between 1.44 and 2.11 inches. In one of the studies, 46 percent of the participants found it necessary to buy larger brassieres at the end of the twelve-week hypnosis intervention. Further, it was found that participants who had better visual imagery skills had the greatest success.[28]

Hypnotic suggestions may also be able to influence the repair of bone fracture. Research psychologist and hypnosis expert Carol Ginandes and radiologist Daniel Rosenthal, both affiliated with Harvard Medical School, have tackled this intriguing question. They recruited twelve people with broken ankles who did not need surgery, and who had been given the standard orthopedic treatment—including serial radiographs and clinical assessments—at Massachusetts General Hospital. Participants were then randomized to either a treatment group or a control group. The treatment group received a hypnotic intervention (individual sessions) once a week for twelve weeks. This intervention was designed to accelerate fracture healing. The casts were applied by the same doctor, and the same radiologists took regular X-rays to monitor how well the participants healed. In addition, the radiologist who examined the X-rays did not know which participants underwent hypnosis.

X-rays revealed an important difference in fracture-edge healing at six weeks following injury, and orthopedic evaluations showed trends toward greater healing for hypnosis participants through week nine, including improved ankle mobility and better functional ability to descend stairs. Ginandes and Rosenthal concluded that hypnosis may be capable of enhancing fracture healing.[29]

Changes in the body resulting from hypnotic suggestions are sometimes unpredictable. Deirdre Barrett, another psychologist and hypnotherapist on the faculty of Harvard Medical School, has reported the case of a man who underwent hypnotherapy to quit smoking. During one hypnosis session, Barrett gave the man the suggestion to imagine himself as the person he would like to be. Rather than imagining himself as a nonsmoker, the man—a homosexual—imagined himself as a pregnant woman. For a long time, apparently, he had wished that he could bear a child. After the suggestion by Barrett, the man continued on a regular basis to visualize himself as a pregnant woman. When he arrived at the hospital three months later, he was not pregnant—

but he had an expanded abdomen and one of his breasts was enlarged. Furthermore, his nipples were secreting and he was suffering from morning nausea.[30]

||||

A few skeptics have attempted to discredit hypnosis by arguing that it is nothing more than an amplified form of social compliance: that is, hypnotic subjects adopt a role to cooperate with the wishes of the hypnotist. One way to test the validity of this view is to investigate whether measurable changes can be detected in the brain of hypnotized people. In 2000 Stephen Kosslyn, a professor of psychology at Harvard University, and his colleagues examined this titillating question. Specifically, they sought to determine whether hypnosis can influence the brain mechanisms underlying color perception.[31]

The researchers pretested 125 individuals. Eight highly hypnotizable people were recruited for the study (this category includes about 8 percent of the general population—about half of the people can be hypnotized to a medium extent). Each participant was hypnotized as he or she lay in a positron emission tomography (PET) scanner at Massachusetts General Hospital. During hypnosis, a computer screen overhead presented patterns of red, blue, yellow, green, and gray rectangles. The PET scanner recorded the participants' brain activity while they were instructed either to "color" the gray rectangles with their minds or to see gray rectangles when they were in fact looking at brightly colored rectangles. Participants were also scanned when they were not hypnotized as a baseline.

The results showed that when participants were hypnotized, color areas of both the left and right hemispheres were activated when they were asked to perceive color, and decreased when they were told to see gray—regardless of which rectangles they were being shown. That is, when the participants *believed* they were looking at color, the parts of their brains involved in color vision showed increased blood flow; and when they believed they were looking at gray, these brain areas showed decreased blood flow.

The results of this PET study cannot be explained by saying that hypnotic subjects are just telling hypnosis researchers what they want to hear. Indeed, these results demonstrate that hypnotized volunteers can change the way the brain processes information.[32]

In recent years, other brain imaging studies have been performed to learn how hypnosis works. One of these studies was carried out by a team of researchers at the University of Iowa. The study was led by anaesthesiologist Sebastian Schulz-Stübner. The researchers used fMRI to evaluate whether hypnosis changes brain activity in a way that might explain how it reduces pain perception.[33]

Twelve healthy participants had a heating device put on their skin to determine the temperature that each of them felt to be painful (8 out of 10 on a 0–10 subjective pain scale). Next, the participants were divided into two groups. One group was hypnotized and placed in the fMRI machine, where their brain activity was recorded while the painful heat was applied. Then the hypnotic state was terminated and another scan was carried out without hypnosis while the same painful thermal stimulation was again applied to their skin. The participants in the second group underwent their first scan without hypnosis. These scans were followed by second scans performed under a hypnotic state.

Under hypnosis, all of the participants experienced a significant pain reduction (less than 3 on the 0–10 self-reported pain scale) in response to the painful heat. Moreover, different patterns of brain activity were seen compared to when the patients were not hypnotized and received the painful stimulation. Specifically, brain activity was reduced in areas of the pain network.

The pain network, as its name suggests, is involved in the perception of pain. It operates like a relay system. An input pain signal from a peripheral nerve goes to the spinal cord, where the information is processed and transmitted to the brainstem. From there the signal is passed on to the midbrain region and finally arrives in the brain area implicated in the conscious perception of painful stimulation. Schulz-Stübner and his colleagues speculated that decreased activity in regions of the pain network may prevent pain signals from reaching the parts of the brain associated with pain perception.

Hypnosis can also be useful for investigating so-called conversion disorder, or what used to be termed *hysteria*. This enigmatic condition involves people who experience symptoms affecting sensory or motor function, such as blindness, numbness, and paralysis, with no identifiable neurologic explanation. Toward the end of the nineteenth century, Sigmund Freud proposed that conversion disorder symptoms are the result of emotional conflicts, which are "converted" into neurological ailments.

Neuroscientist Martin Pyka and fellow researchers at the University of Marburg, Germany, have used hypnosis to explore what may be going on in the brain in conversion hand paralysis.[34] Nineteen highly suggestible participants were recruited in this fMRI study. Participants underwent two sessions of fMRI, one under the hypnotic suggestion of a left-hand paralysis and the other session while in a normal state.

Researchers began hypnotic induction with suggestions such as "The left hand feels weak, heavy." These suggestions were followed by direct instructions like "The left hand is paralyzed, you cannot move the hand anymore."

The left-hand paralysis induced by hypnosis was not associated with activation of brain areas involved in the inhibition of movement but with increased coupling between regions implicated in the representation of the self and those that monitor one's own movements. These results indicate that the hypnotic suggestions did not lead to the actual suppression of the participants' hand control. Rather, it seems that the participants no longer thought they had the power to move their left hands.[35]

||||

Hypnosis can be a powerful tool for harnessing the power of our minds and affecting the way our brains and bodies function. Research shows that suggestions received in a hypnotic state of trance can markedly reduce pain perception and improve various skin conditions, allergies, and asthma. Hypnotic suggestions can also alter the activity of the brain and even lead to spectacular body changes.

As researcher Emily Williams Kelly points out, these bodily effects are much more specific than the general physiological processes associated with simple relaxation.[36] The mechanisms by which specific hypnotic suggestions can trigger specific physical effects and contribute to the healing of the body still remain elusive.

It seems clear that the phenomena apparently produced by suggestions from the hypnotist largely depend on the subject's own mental activity. We are not, in fact, being controlled by hypnotic suggestion; rather, hypnosis can help us let down the normal barriers that prevent us from using the abilities that lay dormant within us. In the hypnotic state, subjects seem to be able to access deeper levels of the mind. These deeper levels allow a connection with a larger intelligence hidden within us, which has a much greater capacity than the normal waking mind to influence what is going on in the body.[37]

In this chapter we have seen that suggestions and images received in a trance state can substantially affect our own body's functioning. But can mind influence the bodily functions of other people at a distance? Can it receive information that is beyond the reach of our ordinary senses?

6

Beyond Space and Time

Psi

There are two ways to be fooled. One is to believe what
isn't true; the other is to refuse to believe what is true.

—PHILOSOPHER SØREN KIERKEGAARD[1]

In 1970, Joe McMoneagle, a bright young Army intelligence officer, was sitting in a restaurant in Brassau, Austria, drinking a
rum and Coke. Suddenly, he began to feel strange, slowed down;
the sights and sounds around him seemed to be changing. He
decided to leave the restaurant and walked toward the door.
"I remember suddenly being intensely frightened by what was
happening to me but hopelessly unable to do anything about it,"
he writes. "My last blurred memory was the door opening and
my body falling through it from its own momentum. I distinctly
remember fearing that I would break the glass with my fall and
then heard a horribly loud pop and thought it might have been
my face striking something as I was falling. Expecting to feel the

cobblestones smack me in the face, I found myself catching my balance and then standing in the street."[2]

Joe felt well again, so he was shocked to look down and see that his body was lying in the street, face bloody, eyes and mouth open. McMoneagle was having a near-death experience, or NDE (we explore NDEs in depth in the next chapter) and was reluctant to return to his body. He later reported that "I wanted to remain in the Light . . . because it felt as if all knowing and feeling were contained there. It was like swimming in nothing but pure and unconditional love."[3]

McMoneagle returned to the Army, but he was fundamentally changed: his NDE had apparently given him the ability to achieve altered states of consciousness at will. Eventually, his abilities caught the attention of the Cold War army, which was looking for promising new ways to compete with the USSR. In 1977, he was one of the first officers recruited for the top-secret Stargate Project backed by the army and other U.S. government agencies including the Central Intelligence Agency (CIA), the Defense Intelligence Agency (DIA), and the navy. These agencies were interested in testing whether remote viewing—the mental faculty that allows a perceiver (or "viewer") to describe or give details about a target that is inaccessible to normal senses due to distance or shielding—represents a truly efficient means to obtain sensitive intelligence information.

McMoneagle's task was to use remote viewing to spy on enemies of the United States, and he became one of the military's most accurate remote viewers. In September 1979, the National Security Council (NSC) asked him to "see" inside a huge building located somewhere in northern Russia. A spy satellite photo had revealed suspect construction activity around the building, and the NSC wanted to know what was going on inside.

Initially, McMoneagle was given only the map coordinates of the building. He described a cold location and a large building near a huge body of water. This information was accurate, so he was then shown the spy photo and asked to describe what was inside the building. McMoneagle reported that the interior was a gigantic working area, full of beams of steel and materials used for constructing scaffolds. He also described blue flashes suggestive of arc welding lights. In another remote viewing session, McMoneagle sensed that an enormous submarine was being constructed in one part of the building. He drew a sketch of what he "viewed": a long, flat deck, a new kind of drive mechanism, a double hull, and angled missile tubes with room for several missiles.

The members of the NSC thought McMoneagle must be mistaken. What he was describing was much larger than any submarine that either the Americans or the Russians had at the time. Four months later, however, spy-satellite photos revealed that the largest submarine ever built was traveling through an artificial channel from the building to the body of water. The photos also showed that the submarine had a large, flat deck and twenty missile tubes.[4]

On December 17, 1981, U.S. Army General James Dozier, deputy chief of staff of NATO's Southern European land forces, was kidnapped by the Red Brigades, an Italian terrorist group. Unable to find the general, the U.S. Army turned to their psychic spies for intelligence. McMoneagle was especially successful. He was able to get the location—Padua, in northern Italy. He was also able to describe the room in which Dozier was being held chained to a wall heater.

At his retirement, Joe McMoneagle received the Legion of Merit for his contribution to various intelligence operations. This military decoration is awarded to members of the armed forces for outstanding conduct in the performance of meritorious service to the United States. His career as a military remote viewer suggests that it is possible for people to receive information from a distance using psi abilities that are beyond the reach of the ordinary senses.

IIIII

British psychologist Robert Thouless coined the term *psi* in 1942 as a neutral term for a wide range of psychic phenomena including extrasensory perception (ESP) and psychokinesis (PK). Joe McMoneagle's psi ability—which is also called clairvoyance—represents one type of ESP. Other forms include telepathy, or "mind reading"—information exchanged between two or more minds, without the use of ordinary senses—and precognition—knowing something will occur before it does occur when the knowledge could not be inferred by ordinary means. Interacting mentally at a distance with animate or inanimate matter represents another generic category of psi phenomena.[5]

Although people throughout history and across all cultures have reported psi experiences, psi phenomena have been ridiculed and rejected as impossible by some scientists and journalists—even though thousands of controlled scientific experiments conducted over the past decades have demonstrated the reality of psi effects. The results of these experiments have been published in reputable scientific journals, including *Foundations of Physics,*

Physical Review, American Psychologist, and *Psychological Bulletin.* In 1995, the U.S. Congress asked the American Institutes for Research to review previously classified government-sponsored psi research for the CIA. One of the main reviewers, statistician Jessica Utts of the University of California–Davis, concluded that "the statistical results of the studies examined are far beyond what is expected by chance. Arguments that these results could be due to methodological flaws in the experiments are soundly refuted." She continued:

> *Effects of similar magnitude to those found in government-sponsored research . . . have been replicated at a number of laboratories across the world. Such consistency cannot be readily explained by claims of flaws or fraud. . . . It is recommended that future experiments focus on understanding how this phenomenon works, and on how to make it as useful as possible. There is little benefit to continuing experiments designed to offer proof.* [6]

Today, although a growing number of scientists are accepting the reality of psi, these phenomena are still considered "anomalies"—unexplained deviations from the norm—because no current theories in physics, psychology, or neuroscience can explain them convincingly. Understanding these phenomena thus represents a stimulating challenge. Anomalies should not simply be regarded as mistakes, inconveniences, or curiosities: the history of science demonstrates that anomalies occasionally lead to major breakthroughs and even scientific revolutions. [7]

Unfortunately, the history of science shows very clearly that not all scientists are the models of open-mindedness, objectivity, and rationality they would hope to be. For instance, in 1772 the French Academy of Science, Europe's leading, rational authority, created a committee to examine reports of "stones falling from the sky." After having examined the evidence and deliberated, Antoine Lavoisier, the father of modern chemistry, and his fellow academicians concluded that "stones cannot fall from the sky, because there are no stones in the sky!"

As a consequence, the concept of what we commonly call meteorites was denounced by the members of the academy as nothing but a delusional absurdity. Farmers who brought to the French Academy samples of meteorites that had fallen in their fields were disdainfully perceived as superstitious, ignorant peasants. But things were about to change. On the night of April 26, 1803, the people of l'Aigle in Normandy were abruptly woken up by a meteorite shower of more than 3,000 fragments. Physicist Jean-Baptiste Biot was sent by the academy to investigate the event. His report led the French Academy of Science to reluctantly admit that stones could indeed fall from the sky.

The reactions to the invention of the first successful airplane also admirably illustrate how scientists are not always guided by rational thought and objectivity. For five years, from December 1903 to September 1908, Wilbur and Orville Wright, two bicycle mechanics from Ohio, repeatedly claimed to have constructed a heavier-than-air flying machine and to have flown it successfully. But despite much evidence—including several public demonstrations, photographs of them flying, and affidavits from local dignitaries—their claim was scorned and dismissed as a hoax by several American scientists. One of them, Simon Newcomb, a professor of mathematics and astronomy at Johns

Hopkins University, was so convinced that powered heavier-than-air flight was absolutely impossible that he rejected the Wright brothers' claim without even bothering to examine the evidence. When President Theodore Roosevelt ordered public trials toward the end of 1908, the Wrights were finally able to prove that they were saying the truth, and the scientists were forced to accept that a flying machine was a reality.[8]

For more than a century, reactions similar to those of Lavoisier and Newcomb have considerably slowed down the progress of psi research. As notes author Chris Carter, this domain of investigation is the only scientific field for which a few "professional pseudoskeptics" systematically attempt to bring the domain's findings into disrepute.[9] These pseudoskeptics want lay people to believe that they are genuine skeptics. But nothing could be further from the truth.

True skeptics conduct an open-minded and objective inquiry for truth. Unprejudiced, they have a questioning attitude toward facts and views, and they are willing to challenge their own beliefs. In contrast, pseudoskeptics are believers, committed to defend scientific materialism. Because psi phenomena demonstrate the falsity of the materialist worldview, pseudoskeptics have no other option than to dismiss all evidence for psi as uncontrolled, unreplicable, or flawed—even if several psi phenomena have been replicated hundreds of times in independent laboratories across the world.

In fact, most pseudoskeptics do not bother to examine the evidence and conduct experiments. The attitude is perfectly epitomized by physicist Hermann von Helmholtz and neuropsychologist Donald Hebb. These two eminent scientists overtly admitted that they were prejudiced, and that no amount of evidence would suffice to convince them of the existence of psi phenomena.

Sometimes, pseudoskeptics do not hesitate to use harsh rhetoric to raise doubts about the competence and integrity of psi researchers. As biologist Rupert Sheldrake points out, some of these pseudoskeptics behave like fanatical fundamentalists engaged in a holy war to defend the materialist doctrine.[10] For these zealots, the materialist credo is tantamount to science and reason. Since humility is not their forte, they pretend to know what is possible and what is not. Again and again, they take their ignorance and their limited beliefs about the world for reality.

Their emotional attachment to the materialist ideology enables pseudoskeptics to easily deny the existence of psi phenomena because these events and experiences do not fit with their preconceived view of the world. In doing so, they avoid being forced to relinquish their deeply held, cherished beliefs. It is ironic that these pseudoskeptics envision themselves as being the champions of rationality.

Throughout this book, we engage in a rational look at the real scientific evidence for "unexplained" phenomena. So, in this chapter, we will look at the scientific evidence for the real existence of psi phenomena, beginning with telepathy.

||||

In the 1970s, three researchers—parapsychologist Charles Honorton at the Maimonides Medical Center in New York, psychologist Adrian Parker at the University of Edinburgh, and William Braud, a psychologist at the University of Houston— independently arrived at the hypothesis that a reduction of sensory stimulation should favor the occurrence of psi. This hypothesis was based on the association between decreased mental "noise"—the constant, involuntary, chatter of the mind—and the manifestation of psi phenomena noted thousands of years ago in the Vedas, the religious texts of ancient India. The Vedas assert that sustained meditation practice can lead to the development of various psi abilities. This claim suggests that when mental "noise" decreases, it becomes easier to attend to and detect psi information.[11]

Honorton, Parker, and Braud surmised that psi operates as a weak signal that is usually concealed in stronger signals related to our senses and continually bombarding us. Separately, these researchers devised a telepathy experiment based on the *ganzfeld condition*—a sensory deprivation technique that rapidly evokes an agreeable, dreamy state of awareness and that was originally developed to investigate visual mental imagery. The three researchers speculated that with sensory stimulation blocked, the likelihood of perceiving subtle impressions would be greatly enhanced.

Ganzfeld telepathy experiments involve two participants: one who concentrates on mentally transmitting an image (the sender) and another who receives it (the receiver). These experiments are generally conducted in three phases: preparation, sending, and judging. In the preparation phase, the receiver sits in the ganzfeld room in a cozy reclining chair, wearing translucent hemispheres (usually halved Ping-Pong balls) over her eyes and headphones on her ears. A red light shines on her face as she listens to a constant stream of white noise played through the headphones. The experimenter asks an assistant to randomly select one "target pack" of pictures out of a large pool of such packs. Each pack comprises four pictures, and all the target packs and target pictures are hidden inside opaque envelopes. The receiver is then locked in the room.

During the sending phase, the experimenter gives the sender the target, still in its opaque envelope, and the sender is then locked in a room. The sender views the target picture and tries to mentally send it to the receiver. The sender is requested to attempt to become "immersed" in the target picture and to send her "full experience."

In the judging phase, the experimenter informs both the sender and the receiver that the sending phase is terminated. The red lamp and the white noise are turned off, and the Ping-Pong ball eyeshades are removed. The experimenter then presents the receiver with copies of the four pictures and asks her to rank these pictures (1 to 4) as to how well each accords with her subjective impressions during the sending phase. A "hit" is scored if the receiver ranks the target number 1. If not, the experiment is considered a "miss." Statistically this experiment has a 25 percent chance hit rate—that is, it should produce a hit every four sessions.[12]

At the beginning of the 1980s, Honorton and his colleagues improved the methodological quality of ganzfeld experiments by creating a fully automated procedure they called the "auto-ganzfeld." In typical autoganzfeld experiments, the target pool consists of eighty short audio-video clips (taken from motion pictures, cartoons, and TV shows). The video-based target pack and the video-clip target are selected randomly by a computer, and a closed-circuit video system presents the target to the sender over a video monitor. Additionally, the interactions among the experimenter, receiver, and sender are totally automated, and a computer presents the four targets in random order to a video monitor in the receiver's room.[13] Honorton and his co-workers carried out a six-year research program using autoganzfeld sessions. Overall, 240 people participated as receivers in 354 autoganzfeld sessions. The hit rate was 37 percent and the odds against chance were 45,000 to one.[14]

Positive autoganzfeld results have been independently replicated by several research psychologists, including Dean Radin of the Institute of Noetic Sciences in northern California, Dick Bierman of the University of Amsterdam, Daryl Bem of Cornell University, and Adrian Parker, who is now at the University of Göteborg in Sweden. A meta-analysis of the replication studies conducted by Radin, Bierman, Bem, and Parker has revealed a hit rate of 33.2 percent with odds against chance beyond a million billion to one.[15]

Meta-analyses combine the results of several studies to more accurately estimate the true magnitude of phenomenon, that is, its "effect size." This form of statistical analysis allows scientists to aggregate several small-scale experiments that often, on their own, cannot reveal weak effects such as psi. One complication in conducting meta-analysis has been called the "file-drawer problem" by Robert Rosenthal, a professor of psychology at the University of California–Riverside.[16] This problem refers to hypothetical studies that scientists performed but did not bother to publish—and consigned to the "file drawer"—because they produced a nil result. For a long time, critics have argued that these hypothetical studies were concealed on purpose by psi researchers.

Fortunately, meta-analyses provide a solid basis for calculating how many file-drawer studies there would have to be to explicate the positive results that have been published. In the case of the experiments conducted by Radin, Bierman, Bem, and Parker, the 7 percent above-chance effect was associated with odds against chance of 29,000,000,000,000,000,000,000 (or 29 quintillion) to one. A conservative assessment of the number of experiments required to invalidate this finding is 2,002.[17] This number represents a ratio of twenty-three file drawer studies to each published experiment. In other words, each of the thirty known investigators would have had to perform but not report sixty-seven supplementary experiments. Considering that the average ganzfeld experiment had thirty-six trials, these 2,002 "concealed" investigations would have required 72,072 additional sessions. To produce this number of sessions would entail unceasingly running ganzfeld sessions 24 hours a day, seven days a week, for 36 years. As Radin points out, this is not credible.[18]

Clearly, then, telepathy does occur in the ganzfeld condition. At this time, more than fifty researchers have reported successful replications from laboratories across the United States, Sweden, United Kingdom, Argentina, Australia, and Italy. Some ask why hit rates are not higher, but it is important to realize that the 32 percent hit rate (a 7 percent effect above chance) was obtained mainly with volunteers who do not claim any particular abilities.[19] When special populations are investigated, such as creative artists, much higher hit rates (for example, 47 percent) are measured.[20]

|||||

Telepathy experiments indicate that information can be transmitted mentally, across space. But is it possible to detect information about an event before that event takes place?

Experimental studies of precognition have been conducted for more than seventy years. In the early experiments, participants were asked to guess which one of various potential targets—ESP card symbols, the faces of a die, or an array of colored lightbulbs—would later be randomly selected. A meta-analysis including 309 precognition studies published between 1935 and 1987 produced odds against chance of 10^{25} (10 million billion) to one. Chance was thus eliminated as a reasonable explanation. Furthermore, the file drawer problem was discarded by determining that the number of unpublished, unsuccessful studies needed to nullify these enormous odds was 14,268.[21]

A few decades ago, researcher Dean Radin pioneered a novel way to study precognition experimentally by exploring whether further emotional states (or future feelings) are detectable in present nervous system activity. Such emotional responses are known as *presentiment*—the vague sense that something is about to occur, but without any conscious awareness of a specific event.[22]

In the presentiment experiments conducted by Radin and his colleagues, participants sat in a comfortable chair placed in front of a computer monitor. Electrodes recording fluctuations in skin conductance—an index of emotional reactivity—were attached to the first and second fingers of their left hands. A sensor was also attached to the pad of the third finger of their left hands to record both heart rate and the amount of blood in the fingertip. Participants held a computer mouse in their right hands and pressed the mouse button when they were ready to begin. This caused the computer to randomly select one target photo out of a large collection of pictures. The computer screen, however, remained blank. After five seconds of the blank screen, the selected photo was displayed for three seconds. Then the computer screen went blank again (for five to ten seconds). This was followed by a five-second rest period. Following the rest period, participants were instructed to press the mouse button whenever they felt ready for the next trial. The three physiological responses were recorded at the same time. In a single session, participants viewed forty pictures. On each trial, the computer randomly chose one target photo from a collection of 120 photographs. One category of pictures elicited calmness (pleasant pictures of landscapes and nature scenes, for example), whereas the other category of pictures induced emotional arousal (such as erotic photos, or autopsies). The physiological measures were averaged for all the "calm" trials and the "emotional" trials.[23]

As expected, viewing the emotional pictures led to increased skin conductance, heart rate deceleration, and blood volume decrease. In addition, the responses to the calm pictures reflected a state of relaxation. Strikingly, the physiological measures revealed that the participants were starting to respond to the emotional pictures *before* these pictures were presented, that is, as soon as the mouse button was pressed. Participants were asked after the experiment if they were consciously aware of the upcoming pictures. Most of them responded no, indicating that presentiment is mostly an unconscious process.

Radin conducted a series of four presentiment experiments. The combined odds against chance for these experiments were 125,000 to 1. These results were replicated independently by Dick Bierman and his colleagues at the University of Amsterdam. In that replication experiment, adult volunteers lay in a fMRI scanner while they looked at computer-projected images. After each picture they were instructed to remain as calm as possible, to not think about the pictures that had already been presented, and to avoid anticipating the forthcoming pictures. On each trial the pictures were selected randomly; no one knew in advance which picture was about to be presented. Remarkably, specific areas of the brain involved in emotion were activated in ten participants *before* erotic pictures appeared. This suggests that the brains of the participants were somehow responding to future events.[24]

There have now been more than forty replications of the presentiment experiment, reported by several laboratories around the world, with physiological measures including skin conductance, heart rate, pupil dilation, and EEG.[25] Taken together, the results of these experiments demonstrate that we can unconsciously perceive our futures.

Another well-designed study showing experimental evidence for precognition has recently been published in the *Journal of Personality and Social Psychology* (*JPSP*). This study—which included nine experiments involving more than 1,000 participants, Cornell undergraduates—was performed by Daryl Bem, an emeritus professor and an acclaimed research psychologist at Cornell University.[26] In one of the experiments, Bem provided the following instructions to 100 participants:

This is an experiment that tests for ESP. It takes about 20 minutes and is run completely by computer. First you will answer a couple of brief questions. Then, on each trial of the experiment, pictures of two curtains will appear on the screen side by side. One of them has a picture behind it; the other has a blank wall behind it. Your task is to click on the curtain that you feel has the picture behind it. The curtain will then open, permitting you to see if you selected the correct curtain. There will be 36 trials in all. Several of the pictures contain explicit erotic images (e.g., couples engaged in nonviolent but explicit consensual sexual acts). If you object to seeing such images, you should not participate in this experiment.[27]

Participants were thus led to believe that they were participating in an experiment whose objective was to test the possibility of clairvoyance. But in reality, neither the pictures themselves nor their left/right position were determined until after the participants recorded their guesses, making the experimental procedure a test of precognition. Since the location of the pictures was selected randomly by the computer, the participants should have correctly guessed the location of the erotic photos 50 percent of the time. Results, however, revealed that the participants performed above chance and correctly located the erotic pictures 53.1 percent of the time, whereas the hit rate on "non-erotic pictures" did not deviate from chance.[28]

The editors of *JPSP* have been severely criticized for the publication of Bem's article. For instance, Ray Hyman, a retired psychology professor and a longtime skeptic of psi phenomena, has characterized the article as "pure craziness . . . an embarrassment for the entire field"; while cognitive scientist Douglas Hofstadter noted that the publication of work like Bem's may contribute to unleashing and legitimizing other "crackpot ideas." Hofstadter further complained that "if any of [Bem's] claims were true, then all of the bases underlying contemporary science would be toppled, and we would have to rethink everything about the nature of the universe." The *JPSP*'s editors responded to these critics that "our obligation as journal editors is not to endorse particular hypotheses, but to advance and stimulate science through a rigorous review process."[29]

What brought on these rude criticisms, considering the broad and compelling implications of Bem's study and other research on precognition? Science writer Jim Schnabel wrote:

> *But how shall we account for the Inquisitional outbursts from scientists that appeared in the [New York] Times . . . ? I mean the calls by prominent academic researchers to effectively suppress the findings of a scientific colleague, the eminent psychologist Daryl Bem, essentially because his findings threatened their reality. . . . Note the absence of scientific reasoning in these statements, and its replacement by fear and loathing. . . . Modern [science's] ideals prohibit it from rejecting ideas just because its elites find them threatening or ontologically untidy. Science is supposed to let the chips fall where they may. As historians and sociologists of science have been pointing out for decades now, it appears to be human nature to want a relatively stable reality, and even scientists will defend their reality instinctively, by fair means or foul. . . . This begs the question of whether academic science is even the place for truly innovative, reality-disturbing research.[30]*

ESP research shows that we can receive information across space or time, without the use of ordinary senses. Can we also mentally influence at a distance inanimate matter and living organisms?

Psi researchers have long been exploring this intriguing question. In the 1960s Helmut Schmidt, a physicist working as a research scientist at the Boeing research laboratory in Seattle, began using random-number generators (RNGs) to explore mental interaction with matter. A random-number generator is an electronic circuit that produces thousands of random coin-flips per second. But rather than heads and tails, the RNG produces sequences of random bits, 0s and 1s. In a prototypical experiment, participants are asked to mentally influence the RNG's output so that it produces more 0s than 1s (or vice versa).[31]

Contemporary RNG devices mainly rely upon one of two random sources, radioactive decay times or electronic noise based on quantum processes. These two sources provide electronic spikes that may occur randomly thousand (or even million) times a second. These spikes are used to generate sequences of random bits by interrupting a precise, crystal-controlled clock. The random bit used (0 or 1) is determined by the state of the clock when it is interrupted by a random spike. Participants in RNG experiments frequently receive feedback about the distribution of random events in the form of computer graphics or sounds. All aspects of modern RNG are fully automated, including the presentation of instructions, the feedback provided on a trial-by-trial basis, as well as data storage and analysis.[32]

In 1987, research psychologist Roger Nelson, of Princeton University, and Dean Radin conducted a meta-analysis of RNG experiments. A total of 597 studies conducted by sixty-eight different researchers were included in their meta-analysis. The overall results generated odds against chance beyond one trillion to one. The number of unreported studies required to render null the statistically significant RNG effect was found to be 54,000.[33]

About a decade later, engineer Robert Jahn and his colleagues at the Princeton Engineering Anomalies Research Laboratory (PEAR Lab) published a review of twelve years of RNG experiments performed in their lab. These experiments included more than a hundred volunteers, who were asked to try to intentionally influence the RNG outputs to drift above the chance-expected average (the *high aim* condition), then the below-chance average (the *low aim* condition). The results showed that when the participants wished for high scores, the RNG outputs drifted up, whereas when the participants wished for low scores, the RNG outputs drifted down. Jahn and his co-workers estimated that the results over the entire database produced odds against chance of 35 trillion to one. Interestingly, in some of the PEAR experiments, the participants were thousands of miles away from the RNG, and no decline in effects was found as a function of distance.[34]

Other psi researchers have conducted studies to investigate whether it is possible to mentally influence living organisms at a distance. In the case of studies involving only humans, physiological measures and a sender–receiver protocol were used. In these studies, the autonomic or central nervous system of a receiver was examined while a remote sender was trying to influence the receiver. One of the pioneers in this kind of investigation is University of California–Davis research psychologist Charles Tart. In 1963, he measured skin conductance, heart rate, and blood volume in a sender–receiver protocol. Acting as the sender, Tart received random electrical shocks to determine whether remote receivers could react to those events. Physiological measures in the receivers showed that they reacted significantly to the remote shocks. They were not consciously aware, however, of the events.[35] In independent studies conducted subsequently, psychologists Erlendur Haraldsson (in Iceland) and Jean Barry (in France), as well as engineer Douglas Dean (at the Newark College of Engineering in New Jersey), recorded significant alterations in receivers' finger blood volume when senders, occasionally located thousands of miles away, mentally directed emotional thoughts and feelings toward them.[36]

Psychologist William Braud and his colleague, anthropologist Marilyn Schlitz, have carried out several experiments related to direct mental interactions with living organisms when they were working at the Mind Science Foundation in San Antonio, Texas. Braud and Schlitz performed a meta-analysis of the thirty-seven experiments they had conducted there until 1991. These experiments, which involved the measurement of various physiological responses, comprised 655 sessions, with 153 people acting as senders and 449 people or animals acting as receivers. The combined experiments resulted in odds against chance of more than 100 trillion to one. In fifteen of these experiments, the sender and the receiver were isolated by distance, and the receiver's skin conductance was constantly monitored. At randomly selected periods, the sender was instructed to attempt to arouse or calm the distant receiver by thinking about that person. During the control periods, which were also randomly selected, the sender was asked to focus her attention elsewhere. The results of these experiments provided evidence for successful remote influence of skin conductance.[37]

These studies strongly support the view that people can respond unconsciously to distant mental influences. Other investigations indicate that people can mentally influence at a distance living organisms, such as enzymes, bacteria, plants, mice, and dogs.

A number of studies have shown that the brains of pairs of separated individuals with a strong emotional bond can sometimes display unusual correspondence. In one of these studies, when one twin was instructed to close his or her eyes, which causes an increase in alpha brain wave activity, the production of alpha waves was also enhanced in the brain of the distant twin. This effect was not found in unrelated pairs of people.[38]

One of the replication studies concerning this phenomenon has been recently conducted by neuroscientist Leanna Standish and her colleagues at Bastyr University.[39] In that study, the members of thirty couples with deep affective connections were separated and placed in rooms several feet away from each other. For each couple, one member was selected as the sender, and the other as the receiver. An EEG recorded electrical activity of the occipital cortex (the visual portion of the brain) in each participant. When the senders were exposed to a flickering light, they attempted to transmit an image or thought about the light to their partners.

Five of the receivers displayed significantly higher brain activity when their partners "transmitted" the images or thoughts. The five couples who had scored significant results were then selected to participate in a comparable experiment, this time with fMRI. In keeping with what had been found using EEG, highly significant increases in brain activity were noted in the receivers' occipital cortex while their distant sending partners were exposed to a flickering light.[40] This increased activity was not detected when the sending partners were not stimulated with the light.

||||

So far, no grand theory has been proposed to explain the various kinds of psi phenomena. But this is not an adequate reason for rejecting a priori the vast bulk of experimental evidence for these phenomena. As a matter of fact, the scientific discovery of new phenomena has often preceded explanatory theories, sometimes by several decades or even centuries. For example, after its official launch in 1899, almost seventy years passed before researchers began to understand how aspirin worked. It is thus crucial to remain open-minded and remember that many phenomena once considered impossible by some scientists are now accepted in mainstream science.

ESP research indicates that the mind can obtain specific, meaningful information in ways that transcend the usual limitations of space and time; RNG studies demonstrate that random processes generated at a quantum level can be influenced by mental intention. Both ESP and micro-PK (the influence of mind on atomic particles or electronic devices) suggest the existence of an interface between the *subjective* psyche and the *objective* physical world.

Classical physics describes the universe as consisting exclusively of discrete, localized particles and objects mutually isolated. But modern (quantum) physics has shown that the universe is fundamentally nonlocal—particles and physical objects that appear to be isolated and separate are in fact deeply interconnected, regardless of distance. Researcher Dean Radin has proposed that psi phenomena may be experiences resulting from this *nonlocality* because we are built out of the same "stuff" as the rest of the universe.[41] Radin foresees the possibility that the basic quantum fabric of reality connects everything, including particles, organisms, minds, and brains. This implies that the whole universe would be a single quantum system. It would also explain why we are now and then able to get information about other people's minds or distant events, and why minds can influence at a distance other minds, physical systems, and biological organisms. Radin also speculates that laboratory-produced psi effects are usually weak because the brain has evolved to focus on objects and events that are directly pertinent to the organism's biological survival.[42]

A number of theoretical physicists have raised the possibility that mind plays a central role in physical reality. For instance, Nobel laureate Eugene Wigner postulated that mind and consciousness are essential to understand quantum physics. In line with this, physicists Helmut Schmidt, Richard Mattuck, and Evan Harris Walker have all proposed that in micro-PK studies, the mind directly influences the outcomes of random quantum events produced by RNGs.

Although some of the principles of quantum physics appear to be compatible with psi, Radin acknowledges that our present understanding of quantum theory is not sufficient to explain psi effects. For example, psychological concepts such as meaning, purpose, attention, intention, and beliefs are critically involved in psi phenomena. For now, quantum physics has nothing to say about such concepts.

Psi phenomena have profound implications for our comprehension of the role of mind and consciousness in the universe. These phenomena suggest that mind plays a fundamental role in nature, and that psyche and the physical world are not radically separated. Psi also gives us a hint that it is time to expand our views about reality.

Joe McMoneagle's remarkable psi abilities resulted from an NDE. In the next chapter we explore verified NDEs and OBEs—out-of-body experiences. Materialist scientists would say that OBEs and NDEs are caused by random electrical and chemical experience in the brain, or are a product of false memory, or any number of plausible explanations. But, as we will see, such beliefs are called into question when it can be demonstrated that higher mental functions (such as perception, self-awareness, memory, thinking) can continue when the brain is severely impaired or no longer functioning.

7

Mind Out of Body

Mind, Brain, and Near-Death Experiences

If you wish to upset the law that all crows are black,
you must not seek to show that no crows are, it is
enough if you prove the single crow to be white.

— PSYCHOLOGIST AND PHILOSOPHER
WILLIAM JAMES[1]

In 1991, Atlanta-based singer and songwriter Pam Reynolds felt extremely dizzy, lost her ability to speak, and had difficulty moving her body. A CAT scan showed that she had a giant artery aneurysm—a grossly swollen blood vessel in the wall of her basilar artery, close to the brain stem.[2] If it burst, which could happen at any moment, it would kill her. But the standard surgery to drain and repair it might kill her too.

With no other options, Pam turned to a last, desperate measure offered by neurosurgeon Robert Spetzler at the Barrow Neurological Institute in Phoenix, Arizona. Dr. Spetzler was a specialist and pioneer in hypothermic cardiac arrest—a daring surgical procedure nicknamed "Operation Standstill." Spetzler would bring Pam's body down to a temperature so low that she would be essentially dead. Her brain would not function, but it would be able to survive longer without oxygen at this temperature. The low temperature would also soften the swollen blood vessels, allowing them to be operated on with less risk of bursting. When the procedure would be complete, the surgical team would bring her back to a normal temperature before irreversible damage set in.

Essentially, Pam agreed to die in order to save her life—and in the process had what is perhaps the most famous case of independent corroboration of out-of-body experience (OBE) perceptions on record. This case is especially important because cardiologist Michael Sabom was able to obtain verification from medical personnel regarding crucial details of the surgical intervention that Pam reported.[3] Here's what happened.

Pam was brought into the operating room at 7:15 A.M., given general anesthesia, and quickly lost conscious awareness. At this point, Spetzler and his team of more than twenty physicians, nurses, and technicians went to work. They lubricated Pam's eyes to prevent drying, and taped them shut. They attached electroencephalography (EEG) electrodes to monitor the electrical activity of her cerebral cortex. They inserted small, molded speakers into her ears and secured them with gauze and tape. The speakers would emit repeated 100-decibel clicks—approximately the noise produced by a speeding express train—eliminating outside sounds and measuring the activity of her brain stem.

At 8:40 A.M., the tray of surgical instruments was uncovered and Robert Spetzler began cutting through Pam's skull with a special surgical saw that produced a noise similar to that of a dental drill. At this moment, Pam later said, she felt herself "pop" out of her body and hover above it, watching as doctors worked on her body.

Although she no longer had use of her eyes and ears, she describes her observations in terms of her senses and perceptions. "I thought the way they had my head shaved was very peculiar," she said. "I expected them to take all of the hair, but they did not."[4] She also described the Midas Rex bone saw ("The saw thing that I hated the sound of looked like an electric toothbrush and it had a dent in it . . ."[5]) and the dental drill sound it made with considerable accuracy.

Meanwhile, Spetzler was removing the outermost membrane of Pamela's brain, cutting it open with scissors. At about the same time, a female cardiac surgeon was attempting to locate the femoral artery in Pam's right groin. Remarkably, Pam later claimed to remember a female voice saying, "We have a problem. Her arteries are too small." And then a male voice: "Try the other side."[6] Medical records confirm this conversation; yet Pam could not have heard them.

The cardiac surgeon was right: Pam's blood vessels were indeed too small to accept the abundant blood flow requested by the cardiopulmonary bypass machine, so at 10:50 A.M., a tube was inserted into Pam's left femoral artery and connected to the cardiopulmonary bypass machine. The warm blood circulated from the artery into the cylinders of the bypass machine, where it was cooled down before being returned to her body. Her body temperature began to fall, and at 11:05 A.M. Pam's heart stopped. Her EEG brain waves flattened into total silence. A few minutes later, her brain stem became totally unresponsive, and her body temperature fell to a sepulchral 60°Fahrenheit. At 11:25 A.M., the team tilted up the head of the operating table, turned off the bypass machine, and drained the blood from her body. Pamela Reynolds was clinically dead.

At this point, Pam's out-of-body adventure transformed into a near-death experience (NDE): she recalls floating out of the operating room and traveling down a tunnel with a light. She saw deceased relatives and friends, including her long-dead grandmother, waiting at the end of this tunnel. She entered the presence of a brilliant, wonderfully warm and loving light, and sensed that her soul was part of God and that everything in existence was created from the light (the breathing of God). But this extraordinary experience ended abruptly, as Reynolds's deceased uncle led her back to her body—a feeling she described as "plunging into a pool of ice."[7]

Meanwhile, in the operating room, the surgery had come to an end. When all the blood had drained from Pam's brain, the aneurysm simply collapsed and Spetzler clipped it off. Soon, the bypass machine was turned on and warm blood was pumped back into her body. As her body temperature started to increase, her brain stem began to respond to the clicking speakers in her ears and the EEG recorded electrical activity in the cortex. The bypass machine was turned off at 12:32 P.M. Pam's life had been restored, and she was taken to the recovery room in stable condition at 2:10 P.M.

Pam's experience while clinically dead seems to have been continuous from the time she "popped" out of her body at the sound of the surgeon's saw until she "plunged" back into her body at the close of the surgery. But did it really happen? The evidence is compelling. Robert Spetzler believed that Pam's observations could not have been based on what she experienced before she was anesthetized: the surgical instruments were covered when she entered the operating room, her eyes were taped shut during the operation, her ears were blocked by the noisy speakers, her heart stopped, her brain waves were flat. "I don't have an explanation for it," Spetzler said in a BBC documentary. "I don't know how it's possible for it to happen."[8]

Pam's experience seems incredible: how could her mind and her consciousness have a verifiable experience that was not mediated by her physical body and brain? Skeptics insist that it is not possible. But evidence based on recent scientific studies on NDE indicates that reports like Pam's are no accident: higher mental capacities can indeed continue when the brain is no longer functional.

█████

Tales of otherworldly experiences have been part of human cultures seemingly forever, but NDEs as such first came to broad public attention in 1975 by way of American psychiatrist and philosopher Raymond Moody's popular book *Life After Life*.[9] He presented more than one hundred case studies of people who experienced vivid mental experiences close to death or during "clinical death"[10] and were subsequently revived to tell the tale. Their experiences were remarkably similar, and Moody coined the term *near-death experience* to refer to this phenomenon. The book was popular and controversial, and scientific investigation of NDEs began soon after its publication with the founding, in 1978, of the International Association for Near Death Studies (IANDS)—the first organization in the world devoted to the scientific study of NDEs and their relationship to mind and consciousness.

NDEs are the vivid, realistic, and often deeply life-changing experiences of men, women, and children who have been physiologically or psychologically close to death. They can be evoked by cardiac arrest and coma caused by brain damage, intoxication, or asphyxia. They can also happen following such events as electrocution, complications from surgery, or severe blood loss during or after a delivery.[11] They can even occur as the result of accidents or illnesses in which individuals genuinely fear they might die. Surveys conducted in the United States and Germany suggest that approximately 4.2 percent of the population has reported an NDE. It has also been estimated that more than 25 million individuals worldwide have had an NDE in the past fifty years.[12]

People from all walks of life and belief systems have this experience. Studies indicate that the experience of an NDE is not influenced by gender, race, socioeconomic status, or level of education. Although NDEs are sometimes presented as religious experiences, this seems to be a matter of individual perception. Furthermore, researchers have found no relationship between religion and the experience of an NDE. That is, it did not matter whether the people recruited in those studies were Catholic, Protestant, Muslim, Hindu, Jewish, Buddhist, atheist, or agnostic.[13]

Although the details differ, NDEs are characterized by a number of core features.[14] Perhaps the most vivid is the OBE: the sense of having left one's body and watching events going on around one's body and, occasionally, at some distant physical location. During OBEs, near-death experiencers (NDErs) are often astonished to discover that they have retained consciousness, perception, lucid thinking, memory, emotions, and their sense of personal identity. If anything, these processes are heightened: thinking is vivid; hearing is sharp; and vision can extend to 360 degrees. NDErs claim that without physical bodies, they are able to penetrate walls and doors and project themselves wherever they want. They frequently report the ability to read people's thoughts.

The OBE is quite important from a scientific point of view because it is the only feature of the NDE that can be independently corroborated. In Pam Reynolds's case, for example, her memory of hearing the cardiac surgeon discussing her arteries was corroborated by the medical records.

All the other features of NDEs, however, are based solely on subjective accounts. NDErs often report feelings of peace and joy, and passage though a region of darkness or a dark tunnel often followed by the appearance of an unusually bright light. NDErs also commonly speak of entering an otherworldly realm of unbelievable beauty, where they may hear incredibly beautiful ethereal music or see magnificent gardens or stunning cities. They also commonly speak of encountering deceased relatives and friends, who look younger and healthier than remembered.

NDErs regularly report meeting a "being of light" that radiates completes acceptance and unconditional love and who may communicate telepathically. The results of cross-cultural studies suggest that personal religious or cultural views influence how the experience is described and interpreted.[15] For instance, a Buddhist may believe that the "being of light" encountered during an NDE is Buddha, whereas a Christian may believe that this "being" is Jesus. In the presence of such a being, NDErs oftentimes experience a life review, during which they "see their life flash before them" and relive both major and incidental events, sometimes from the perspective of the other people involved. From this they come to certain conclusions about their life and what changes are needed.[16]

NDErs may also encounter a barrier, such as a wall, a river, or a gate. At this point, they become aware that once they cross this border they will not be able to return to their bodies and resume their lives. Deceased relatives, or the being of light, may explain that the NDEr must return to his or her body. Often, this is related to fulfilling a purpose in life, such as taking care of young children. It is usually at this point that the NDEr is forced to return to his sick or injured body.

Like Pam Reynolds's "plunge into ice water," most NDErs find this forced return to be a very unpleasant experience. In 1967, Reinée Pasarow, who temporarily stopped breathing as the result of a massive allergic reaction, experienced the return to her body as being "imprisoned" in a "foreign substance."

Reinée's story is fascinating for another reason: in the course of her NDE, the seventeen-year-old seems to have experienced all of the other core features I have just described. This is her experience, in her own words:

> *After my mother and I had eaten dinner, an old friend I had not seen for some time dropped by unexpectedly. I was rather embarrassed, because I had been covered with welts and hives for two days as a result of the allergy and looked somewhat grotesque. The swelling became substantially worse, and I had great difficulty in breathing. . . . An ambulance was called, but, as none was soon available in our rural district, two fire trucks responded in the meantime.*
>
> *I was unconscious on the sidewalk in front of our residence, although I was aware of making a tremendous effort to keep*

breathing. Several firemen were working on me when at last the struggle to keep fighting for my life became too tremendous. I stopped breathing and felt a great relief to be free of the burden of trying to stay alive. I slipped into the dark of a totally unconscious but peaceful realm.

Suddenly, I found myself a few feet outside my body, watching with great curiosity as the firemen gave me mouth-to-mouth resuscitation and violently slapped my legs. I remember them thinking that if they could just get me mad enough, I might come back. My mother was splashing water on my ashen face, and the eldest fireman who was giving me mouth-to-mouth resuscitation kept pleading with me mentally not to leave and seeing his own teenaged daughter in his mind's eye.

Just as suddenly, I found myself viewing this cosmically comic scene from slightly above the telephone wires. I saw a young neighbor boy come out of his house upon awakening from his nap, and I tried screaming at his mother to go and get him before he saw all this. Just as I screamed at her, she looked up the driveway and saw him, and my mother said there was nothing she could do, so she [had] best get her child. One of the firemen commented with a great sigh of failure that I had been without a pulse for three minutes.

I felt a pang of guilt that this poor fireman should feel a failure in my death. He was especially touched because I resembled his own daughter. My mother was dazed, hopelessly without any control over the situation and her shock numbed the onset of grief. I remember saying a prayer for her, in hopes that this would help to see her through the pain, but then I realized that she would come to deal with the situation. I wanted to cry out to them all, my mother, friends and the firemen, that everything

*was as it should be, that I was fine. I was telepathically aware
of everyone's feelings and thoughts, and this seemed a burden, as
their pain was as it should be.*

*Delighted at my newly found freedom, I began to soar. I had
become the phoenix, released at last from the limitations of the
physical world. I was exhilarated. Everywhere around me there
was music; the ether of my new universe was love, a love so
pure and selfless that I only longed for more. I became aware of
my favorite uncle's presence: we gleefully recognized each other
although we were now in energy, rather than a physical form.
He travelled with me for a short time, expressing even more love
and acceptance. As a vast light became visible in this sea of light,
however, I was magnetically drawn into it. The closer I got to
this light (closeness, however not meant in the physical sense) the
more love and ecstacy were mine to experience.*

*Finally, I was sucked into the light source, not unlike one is
swept up in a whirlpool. I became one with the light. As I became one with this omnipresent light, its knowledge became my
knowledge. I was in a single instant what my life had been and
what had been of meaning in my life. . . . The superficial aspects
of my life, what I had accomplished, owned and known, were
consumed in that same instant by the energy of the light. However, those acts in which I selflessly expressed love or concern for
my fellow men were glorified and permanently inscribed . . . ,
with total disregard for however humble those moments had
been. . . .*

*Suddenly, I was ejected from the light to the other side of this
new universe, where I realized I would have to make my way.
I recall someone beckoning to me, although the identity of that
person still remains a mystery, for also at that moment it was*

revealed to me that my moment in the cosmic dance was not completed, that there was something for my human race that I must achieve on the physical plane of existence. Coinciding with the moment of that revelation, the light, the universe, or God himself proclaimed IT IS NOT TIME, and that proclamation hurled me from this magnificent universe of love.

I was pushed through a tremendous tunnel of light, through a progressive rainbow of the wavelengths of color, and catapulted back into the physical realm. It was as if the whole process was not just initiated by the proclamation, but was the proclamation IT IS NOT TIME itself.

I found myself, griefstricken and heartsick, again a few feet from my body. I felt as if I had been cast out of paradise, an Eve no longer in the Garden of Eden. The physical realm was coarse and confusing, divided and foreign. A sense of time and space was clamped down upon my being, casting upon my soul a sense of imprisonment and degradation unlike any I had ever known.

The ambulance had arrived, and the attendants were checking for my absent pulse, which still eluded them. I tried to merge again with the body that was once mine but which now seemed like a foreign substance. This required a tremendous effort on my part, and the attendants placing me in the back of the ambulance only made the merging that more difficult. I hovered over my body in the ambulance, and for a brief instant rejoined it. I felt the surge of blood through my veins, and the attendant motioned to the driver that he had a pulse.

The pain of the physical was too much for me to stand, however, and I separated from the body again, hovering both inside and outside the moving ambulance. I watched as the young attendant in the back mouthed DOA (dead on arrival) to the driver about ten minutes into the drive. My mother's pain at this announcement became my pain, and I was angered at the callousness of the ambulance attendant.

I continued watching from several feet above my body as I was wheeled into the emergency room and the first young doctor was unable to revive me again. . . . At that moment, my personal physician burst into the emergency room in his tuxedo, bag in hand.

"Where is she?" he demanded.

"She was DOA," the young doctor announced. . . .

"The hell she is!" shouted my doctor, a family friend of many years, and got down to the business of determining how many shots of adrenaline I had been given. He ordered that I be given up to six large injections of adrenaline, something the other doctors and nurses obviously considered very dangerous. He proceeded to pump me full of adrenaline and give heart massage until at last a pulse was perceived. It is interesting to note that I was fully aware of what was happening both physically and in the minds of those in the emergency room until I was revived, at which point I was very confused.

To the best of my knowledge, I was without a heartbeat approximately fifteen minutes. The incident left me with some minimal brain damage, the effects of which have been totally overcome, although to this day my reflexes reflect the damage.[17]

▐▐▐

The effects of NDEs are intense, overwhelming, and real. A number of studies conducted in United States, Western European countries, and Australia have shown that most NDErs are profoundly and positively transformed by the experience. One woman says, "I was completely altered after the accident. I was another person, according to those who lived near me. I was happy, laughing, appreciated little things, joked, smiled a lot, became friends with everyone . . . so completely different than I was before!"[18]

After the NDE, however different their personalities were before the NDE, experiencers tend to share a similar psychological profile. Indeed, their beliefs, values, behaviors, and worldviews seem quite comparable afterward. These psychological and behavioral changes are not the kind of changes one would expect if this experience were a hallucination.[19] And, as noted NDE researcher Pim van Lommel and his colleagues have demonstrated, these changes become more apparent with the passage of time.[20] Let's take a quick look at the various types of changes reported following NDEs.[21]

Most NDErs come back with a renewed appreciation for life and sense of purpose, an enhanced sense of wonder and gratitude for life itself. They are more capable of enjoying the here and now. They also feel that life is meaningful and purposeful. Often they come to realize that their main goal in life is to find and fulfill their mission.

NDErs frequently return with a changed self-image. In addition, they generally feel greater self-worth and self-confidence and become less dependent on the approval of others.

One of the most remarkable and consistent changes following an NDE is an increased compassion for other people. NDErs are now more capable of being of service to others. They are also more tolerant, more forgiving, and less critical of others. Experiencers spend more time with relatives and friends, and are more likely to donate to charities or to dedicate themselves to a humanitarian cause.

As NDErs become sensitive to the health of Earth's ecosystem, their concerns commonly extend to all forms of life. They experience a reverence for life they may not have felt before.

NDErs regularly report a decline in religious affiliation but an increase in spirituality, with a greater interest in contemplative practices such as prayer, meditation, and surrender. Additionally, NDErs tend to assert that now they know with deep certitude that God exists, regardless of what they had believed before, and that life goes on after the death of the physical body.

Some NDErs yearn for knowledge. These individuals often attempt to recapture some of the knowledge they acquired during their experience, and lost on return to their bodies.

NDErs tend to perceive a life focused on materialistic pursuits as futile and empty. Many say that they can no longer compete with others for material successes. The goal of becoming eminent and powerful does not matter anymore.

NDErs commonly return with a whole gamut of psi abilities. These may include telepathy, clairvoyance, precognition, and spontaneous OBEs.

If these transformational experiences sound worth dying for, you should be aware that NDEs do not have only positive consequences. Many NDErs are deeply shaken and struggle to accept and integrate their newly discovered insights. This is particularly true when their new insights elicit negative reactions and disbelief from family, friends, and health care professionals. In the first few years after the NDE, several experiencers report feeling depressed, lonely, and nostalgic.[22]

A minority of reported NDEs are disagreeable or frightening.[23] They may, for example, feel out of control as they rush through the dark tunnel; or they may experience an absolute void. Each NDE is unique, and the number of features reported by NDErs varies considerably.

IIII

Some skeptics legitimately argue that the main problem with reports of OBE perceptions is that they often rest uniquely on the NDEr's testimony—there is no independent corroboration. From a scientific perspective, such self-reports remain inconclusive.[24] But during the past few decades, some self-reports of NDErs, such as that of Pam Reynolds, have been independently corroborated by witnesses. One of the best known of these *corroborated veridical NDE perceptions*—perceptions that can be proven to coincide with reality—is the experience of a woman named Maria, whose case was first documented by her critical-care social worker, Kimberly Clark.[25]

Maria was a migrant worker who had a severe heart attack while visiting friends in Seattle. She was rushed to Harborview Hospital and placed in the coronary care unit. A few days later, she had a cardiac arrest but was rapidly resuscitated. The following day, Clark visited her. Maria told Clark that during her cardiac arrest she was able to look down from the ceiling and watch the medical team at work on her body. At one point in this experience, said Maria, she found herself outside the hospital and spotted a tennis shoe on the ledge of the north side of the third floor of the building. She was able to provide several details regarding its appearance, including the observation that one of its laces was stuck underneath the heel, and the little toe area was worn. Maria wanted to know for sure whether she had "really" seen that shoe and begged Clark to try to locate it.

Quite skeptical, Clark went to the location described by Maria—and found the tennis shoe. The details that Maria had recounted could not be discerned from the window of her hospital room. But upon retrieval of the shoe, Clark confirmed Maria's observations. "The only way she could have had such a perspective," said Clark, "was if she had been floating right outside and at very close range to the tennis shoe. I retrieved the shoe and brought it back to Maria; it was very concrete evidence for me."[26]

This case is particularly impressive given that during cardiac arrest, the flow of blood to the brain is interrupted. When this happens, the brain's electrical activity (as measured with EEG) disappears after ten to twenty seconds.[27] In this state, a patient is deeply comatose. Because the brain structures mediating higher mental functions are severely impaired, such patients are expected to have no clear and lucid mental experiences that will be remembered.[28] Nonetheless, studies conducted in the Netherlands,[29] United Kingdom,[30] and United States[31] have revealed that approximately 15 percent of cardiac arrest survivors do report some recollection from the time when they were clinically dead. These studies indicate that consciousness, perceptions, thoughts, and feelings can be experienced during a period when the brain shows no measurable activity.

Another example of a corroborated veridical NDE perception is that of Al Sullivan.[32] In 1988, this fifty-six-year-old van driver went to Hartford Hospital with an irregular heartbeat. While Sullivan was being examined, one of his coronary arteries became obstructed. He was immediately rushed into the operation room for emergency coronary bypass surgery, during which he felt himself leaving his body and moving upward. He looked down and saw himself lying on a table, with his chest cut open. He also saw his cardiothoracic surgeon, Hiroyoshi Takata, "flapping his elbows" as if trying to fly.

Following the operation, Sullivan mentioned this odd observation to his cardiologist, Anthony LaSala, who confirmed that he, too, had observed Takata doing this. Nine years later, Takata told preeminent NDE researcher Bruce Greyson that it was indeed his regular habit to instruct medical personnel by pointing with his elbows so as not to touch anything in the operating room until the actual surgery.

Another corroborated veridical NDE perception, by a coronary care unit nurse, occurred during the pilot phase of the study carried out by Pim van Lommel and his colleagues in the Netherlands. The nurse recalls:

During a night shift an ambulance brings in a 44-year-old cyanotic, comatose man into the coronary care unit. He had been found about an hour before in a meadow by passers-by. After admission, he receives artificial respiration without intubation, while heart massage and defibrillation are also applied. When we want to intubate the patient, he turns out to have dentures in his mouth. I remove these upper dentures and put them onto the "crash cart." Meanwhile, we continue extensive CPR. After about an hour and a half the patient has sufficient heart rhythm and blood pressure, but he is still ventilated and intubated, and he is still comatose. He is transferred to the intensive care unit to continue the necessary artificial respiration. Only after more than a week do I meet again with the patient,

*who is by now back on the cardiac ward. I distribute his medi-
cation. The moment he sees me he says: "Oh, that nurse knows
where my dentures are." I am very surprised. Then he eluci-
dates: "Yes, you were there when I was brought into hospital
and you took my dentures out of my mouth and put them onto
that cart, it had all these bottles on it and there was this sliding
drawer underneath and there you put my teeth." I was especially
amazed because I remembered this happening while the man
was in deep coma and in the process of CPR. When I asked
further, it appeared the man had seen himself lying in bed, that
he had perceived from above how nurses and doctors had been
busy with CPR. He was also able to describe correctly and in
detail the small room in which he had been resuscitated as well
as the appearance of those present like myself. At the time that
he observed the situation he had been very much afraid that we
would stop CPR and that he would die. And it is true that we
had been very negative about the patient's prognosis due to his
very poor medical condition when admitted. The patient tells me
that he desperately and unsuccessfully tried to make it clear to us
that he was still alive and that we should continue CPR. He is
deeply impressed by his experience and says he is no longer afraid
of death. 4 weeks later he left hospital as a healthy man."* [33]

||||

NDEs experienced by people who do not have sight in everyday
life are quite intriguing. In 1994, researchers Kenneth Ring and
Sharon Cooper undertook a search for cases of NDE-based perception in the blind. They reasoned that such cases would represent the ultimate demonstration of veridical perceptions during
NDEs. If a blind person was able to report on verifiable events

that took place when he or she was clinically dead, that would mean something real was occurring. Ring and Cooper interviewed thirty-one individuals, of whom fourteen were blind from birth. Twenty-one of the participants had had NDEs; the others had had OBEs only. Strikingly, the experiences they reported conform to the classic NDE pattern, whether they were born blind or lost their sight in later life. The results of the study were published in 1997.[34]

One of the most enthralling cases is that of a forty-one-year-old woman named Nancy, who had had a disastrous experience while undergoing a biopsy related to a possible cancerous chest tumor. During the surgical procedure, the surgeon accidentally cut her superior vena cava, a large vein located in the upper chest. Panicked, the surgeon sewed the vein shut—an additional error that resulted in a blockage of blood flow.

When Nancy woke up in the recovery room shortly after surgery, she started screaming, "I'm blind, I'm blind!"[35] Her face and body were swollen and purple, and she stopped breathing and lost consciousness. Attendants strapped a respirator over her nose and mouth, and wheeled her out of the recovery room on a gurney to have an angiogram that would measure blood flow through the vein. In their haste, however, the attendants accidentally slammed Nancy's gurney into a closed elevator door. At that point her NDE began. She first felt stepping out of herself on the gurney. In the OBE state, she looked down the hall about fifteen or twenty feet and "saw" two men standing there. One of them was her son's father. The other man was her lover at the time. Then Nancy remembers moving toward a beautiful white light and entering the light. After a while, she felt urged to come back in her body.[36]

In Nancy's medical records, Ring and Cooper found confirmation of her surgical misfortune. To corroborate her claims, they interviewed the two men Nancy claimed to have seen standing in the hall while her body was lying on the gurney and after she had lost her sight. The account given by her child's father fitted globally with Nancy's; but her lover, Leon, independently confirmed all of the essential details of Nancy's story. Ring and Cooper also consulted an ophthalmologist, who agreed that the blockage of the superior vena cava could rapidly damage the optic nerve and lead to blindness.

After a thorough investigation of this case and a review of all relevant information, these researchers concluded that there was no possibility for Nancy to see these events with her physical eyes—which, in any case, were almost certainly sightless at that time. Yet all of the evidence indicates that she really did see.

Kenneth Ring also reported another intriguing case of apparent veridical perception by a woman called Anna, who was virtually blind at the time of her NDE.[37] This case was communicated to Ring by Ingegard Bergström, a Swedish NDE researcher. Anna had the capacity only to distinguish light and darkness. She was not able to recognize silhouettes or walk in dimly lit corridors. One day, while sitting in the kitchen by her kitchen table, she had a cardiac arrest. At that moment, she suddenly saw the sink and a stack of dirty dishes, something that would have not been possible for her to see.

Anna said nothing at the time, but made that statement to Bergström in the presence of her husband, who reacted with surprise. Why had she not told him about this before? Because, she replied matter-of-factly, "You never ask me if I saw anything at the time my heart stopped." She then told Bergström that unwashed dishes were piled up in the sink. In the written report Bergström provided to Ring, the husband was said to have appeared quite guilty, as it had been his responsibility to clean the dishes and put them away.

Based on all the cases they investigated, Ring and Cooper concluded that what happens during an NDE affords another perspective to perceive reality that does not depend on the senses of the physical body. They proposed to call this other mode of perception *mindsight*.[38]

||||

Despite corroborated reports, many materialist scientists cling to the notion that OBE and NDE experiences are located in the brain. In 2002, neurologist Olaf Blanke and colleagues at the University Hospitals of Geneva and Lausanne in Switzerland described in the prestigious scientific journal *Nature* the strange occurrence that happened to a forty-three-year-old female patient with epilepsy.[39] Because her seizures could not be controlled by medication alone, neurosurgery was being considered as the next step. The researchers implanted electrodes in her right temporal lobe to provide information about the localization and extent of the epileptogenic zone—the area of the brain that was causing the seizures—which had to be surgically removed. Other electrodes were implanted to identify and localize, by means of electrical stimulation, the areas of the brain that—if removed—would result in loss of sensory capacities or linguistic ability, or even paralysis. Such a procedure is particularly critical to spare important brain areas that are adjacent to the epileptogenic zone.

When they stimulated the angular gyrus—a region of the brain in the parietal lobe that is thought to integrate sensory information related to vision, touch, and balance to give us a perception of our own bodies—the patient reported seeing herself "lying in bed, from above, but I only see my legs and lower trunk." She described herself as "floating" near the ceiling. She also reported seeing her legs "becoming shorter."

The article received global press coverage and created quite a commotion. The editors of *Nature* went so far as to declare triumphantly that as result of this one study—which involved only one patient—the part of the brain that can induce OBEs had been located.

"It's another blow against those who believe that the mind and spirit are somehow separate from the brain," said Michael Shermer, director of the Skeptics Society, which seeks to debunk all kinds of paranormal claims. "In reality, all experience is derived from the brain." [40]

In another article, published in 2004, Blanke and co-workers described six patients, of whom three had atypical and incomplete OBEs.[41] Four patients reported an autoscopy—that is, they saw their own double from the vantage point of their own body. In this paper the researchers describe an OBE as a temporary dysfunction of the junction of the temporal and parietal cortex. But, as Pim van Lommel noted, the abnormal bodily experiences described by Blanke and colleagues entail a false sense of reality.[42] Typical OBEs, in contrast, implicate a verifiable perception (from a position above or outside the body) of events such as their own resuscitation or traffic accident and the surroundings in which the events took place. Along the same lines, psychiatrist Bruce Greyson of the University of Virginia commented that "we cannot assume from the fact that electrical stimulation of the brain can induce OBE-like illusions that all OBEs are therefore illusions."[43]

Materialistic scientists have proposed a number of physiological explanations to account for the various features of NDEs. British psychologist Susan Blackmore has propounded the "dying brain" hypothesis,[44] which states that a lack of oxygen (or anoxia) during the dying process might induce abnormal firing of neurons in brain areas responsible for vision, and that such an abnormal firing would lead to the illusion of seeing a bright light at the end of a dark tunnel.

Would it? Van Lommel and colleagues objected that if anoxia plays a central role in the production of NDEs, most cardiac arrest patients would report NDEs.[45] Studies show that this is clearly not the case. Another problem with this view is that reports of a tunnel are absent from several accounts of NDErs. As pointed out by renowned NDE researcher Sam Parnia, some individuals have reported NDEs when they had not been terminally ill and so would have had normal levels of oxygen in their brains.[46]

Parnia raises another problem: when oxygen levels decrease markedly, patients whose lungs or hearts do not work properly experience an "acute confusional state," during which they are highly confused and agitated and have little or no memory recall. In stark contrast, during NDEs people experience lucid consciousness, well-structured thought processes, and clear reasoning. They also have an excellent memory of the NDE, which usually stays with them for several decades. In other respects, Parnia argues that if the dying brain hypothesis was correct, then the illusion of seeing a light and tunnel would progressively develop as the patient's blood oxygen level drops. Medical observations, however, indicate that patients with low oxygen levels do not report seeing a light, a tunnel, or any of the common features of an NDE.

During the 1990s, more research indicated that the anoxia theory of NDEs was on the wrong track. James Whinnery, a chemistry professor with West Texas A&M, was involved with studies simulating the extreme conditions that can occur during aerial combat maneuvers. In these studies, fighter pilots were subjected to extreme gravitational forces in a giant centrifuge. Such rapid acceleration decreases blood flow and, consequently, delivery of oxygen to the brain. In so doing, it induced brief periods of unconsciousness that Whinnery called "dreamlets."[47] Whinnery hypothesized that although some of the core features of NDEs are found during dreamlets, their main characteristics are impaired memory for events just prior to the onset of unconsciousness, confusion, and disorientation upon awakening. These symptoms are not typically associated with NDEs. In addition, life transformations are never reported following dreamlets.

So, if the "dying brain" is not responsible for NDEs, could they simply be hallucinations? In my opinion, the answer is no. Let's look at the example of hallucinations that can result from ingesting ketamine, a veterinary drug that is sometimes used recreationally, and often at great cost to the user.

At small doses, the anesthetic agent ketamine can induce hallucinations and feelings of being out of the body. Ketamine is thought to act primarily by inhibiting N-Methyl-D-aspartic acid (NMDA) receptors that normally open in response to binding of glutamate, the most abundant excitatory chemical messenger in the human brain. Psychiatrist Karl Jansen[48] has speculated that the blockade of NMDA receptors may induce an NDE. But ketamine experiences are often frightening, producing weird images, and most ketamine users realize that the experiences produced by this drug are illusory. In contrast, NDErs are strongly convinced of the reality of what they experienced. Furthermore, many of the central features of NDEs are not reported with ketamine. That being said, we cannot rule out that the blockade of NMDA receptors may be involved in some NDEs.

Neuroscientist Michael Persinger has claimed that he and his colleagues have produced all the major features of the NDE by using weak transcranial magnetic stimulation (TMS) of the temporal lobes.[49] Persinger's work is based on the premise that abnormal activity in the temporal lobe may trigger an NDE. A review of the literature on epilepsy, however, indicates that the classical features of NDEs are not associated with epileptic seizures located in the temporal lobes. Moreover, as Bruce Greyson and his collaborators have correctly emphasized, the experiences reported by participants in Persinger's TMS studies bear little resemblance with the typical features of NDEs.[50]

IIII

The scientific NDE studies performed over the past decades indicate that heightened mental functions can be experienced independently of the body at a time when brain activity is greatly impaired or seemingly absent (during cardiac arrest). Some of these studies demonstrate that blind people can have veridical perceptions during OBEs associated with an NDE. Other investigations show that NDEs often result in deep psychological and spiritual changes.

These findings strongly challenge the mainstream neuro-scientific view that mind and consciousness result solely from brain activity. As we have seen, such a view fails to account for how NDErs can experience—while their hearts are stopped—vivid and complex thoughts, and acquire veridical information about objects or events remote from their bodies.

NDE studies also suggest that after physical death, mind and consciousness may continue in a transcendent level of reality that is normally not accessible to our senses and awareness. This view is utterly incompatible with the belief of many materialists that the material world is the only reality.

The reaction of scientists toward NDEs is to a large extent influenced by their views about religion and the afterlife. An article published in *Nature* in 1998 revealed that 93 percent of the most prominent and influential scientists in the United States (all members of the National Academy of Sciences) consider themselves nonreligious or nonspiritual and rejected the possibility of an afterlife. This survey can help us understand why these scientists so vigorously refuse to accept the implications of the research on NDEs.[51]

In the next chapter, we go deeper into the realm of mind beyond brain, exploring life-changing transcendent experiences that take us outside ourselves and bring us in touch with the fabric of the universe itself.

8

Embracing a Greater Self

Mystical Experiences

If the doors of perception were cleansed everything
would appear to man as it is, infinite.

—POET WILLIAM BLAKE[1]

One day in 1600, a German shoemaker named Jacob Boehme focused on the exquisite beauty of a reflected beam of sunlight in a burnished pewter dish and fell into an ecstasy in which it appeared to him that he could see into the deepest foundations and the very heart of things.[2] The shoemaker, profoundly transformed, would later become an influential mystic and theologian, a leader of the Protestant revolution that swept Europe.

In March 1969, American astronaut Rusty Schweickart was piloting *Apollo 9*'s lunar module as it orbited the earth. As he gazed down on the tiny blue and green planet, crossing geographical and political frontiers again and again and again, he experienced a shift in consciousness that has transformed his life. "When you go around the earth in an hour and a half," he explained, "you begin to recognize that your identity is with that whole thing. That makes a change. . . ."[3] Profoundly transformed, Schweickart now spends much of his time sharing this experience with people and has become an internationally recognized peace and environmental activist.

In 1976, Allan Smith was thirty-eight, a physician and biomedical researcher and self-identified atheistic materialistic scientist. One evening Smith was alone in his Oakland, California, home, watching a beautiful sunset. He began to notice that the light in both the room and the sky was becoming extremely bright, but not unpleasant at all. At the same time, he noticed that he was feeling "elated," and time slowed down until "only the present moment existed. . . . There was no separation between myself and the rest of the universe." He emerged from this experience about twenty minutes later. "I knew that my life would never be the same," he writes; and indeed, it was not. Although he had received a national prize for his research, Smith left his secure and successful university faculty position to work as a part-time freelance clinician: he needed now to give himself time to explore issues related to consciousness, spirituality, and mysticism, and to integrate his experience into his life.[4]

▮▮▮▮

What happened to Jacob Boehme in his workshop, Rusty Schweickart during the *Apollo 9* mission and Allan Smith in his living room is called a *mystical experience,* or ME. Although they are not inherently religious, MEs constitute the basis of most religious traditions. The history of the world has been considerably influenced by the MEs of great spiritual figures such as Moses and Mohammed. Perhaps for this reason, as we will see later in this chapter, some materialist scientists seek to locate this profound experience in the brain. I believe that this intense, overwhelming, and life-changing experience cannot be simply reduced to cerebral "fritzing."

MEs are characterized by an expansion of consciousness far beyond the usual boundaries of bodies and egos, and beyond the everyday concept of space and time.[5] Their origins are something of a mystery. They can result from regular spiritual practices, such as prayer and contemplation. MEs can also be occasioned by near-death experiences (NDEs), shamanic rituals, fasting, sleep deprivation, and the ingestion of mind-altering substances. Sometimes, they simple occur without any apparent reason, as was Allan Smith's experience. MEs are investigated by researchers in transpersonal psychology. This branch of psychology, which emerged in the late 1960s, recognizes that MEs can provide important insights about the nature of reality.

Walter Stace, a British civil servant and philosopher, greatly contributed to the revival of academic interest in mysticism during the second half of the twentieth century. Stace proposed these common features of MEs: the perception of being one infinite, unbroken life, encompassing all things; feelings of peace, bliss, and joy; the impression of having touched the ultimate ground of reality (sometimes identified as "God"); and a transcendence of space and time.[6] Because MEs are subjective and directly experienced, they are very difficult to communicate to others.

Stace also distinguished between extrovertive and introvertive forms of MEs. In extrovertive MEs, mundane objects, art, music, or nature continue to be perceived through the physical senses but they are now transfigured by awareness of the Unity that shines through them. The case of the German shoemaker Jacob Boehme represents a good illustration of this type of ME.

In the introvertive forms of ME—as experienced by Smith—the ordinary "little self" vanishes momentarily and returns transformed. There is a temporary union, or identity, with the All, which may variously be identified as Brahman, the Universal Self, the One, the Absolute, the Ground of Being, the Source, the Tao, Allah, or God. In this experience, distinctions between the experiencer and other people are abolished. There is also a feeling of oneness with all things in the entire universe, and a realization that the Ground of Being is at the origin of all life and has an infinite number of distinct expressions. In Hinduism, the Mandukya Upanishad refers to this unitary form of consciousness as Turiya. In the West, Canadian psychiatrist Richard Maurice Bucke has coined the term Cosmic Consciousness to designate this most dramatic type of ME.

In 1872, Bucke experienced a few moments of this state of consciousness. Following his experience, he devoted thirty years of his life to writing a book entitled *Cosmic Consciousness*,[7] in which he provides several examples of individuals who he felt had reached a state of Cosmic Consciousness, either temporarily or permanently. These enlightened figures include poets Henry David Thoreau and Walt Whitman, as well as the religious figures Moses, Jesus, and Buddha. Bucke proposed that Cosmic Consciousness is the next stage in the development of human consciousness—the preceding two stages being the simple consciousness of animals and the self-consciousness of the mass of humanity. Bucke also speculated that the frequency of individuals experiencing Cosmic Consciousness has been increasing over the past few thousand years.

Allan Smith later wrote about his experience. As a scientist, Smith was a keen observer of his own experience. His words present a wealth of detail and evocative descriptions of an essentially indescribable experience that began unexpectedly while he was sitting in an easy chair watching the sunset:

I . . . noticed that the level of light in the room as well as that of the sky outside seemed to be increasing slowly. . . . I began to feel very good, then still better, then elated. While this was happening, the passage of time seemed to become slower and slower. The brightness, mood-elevation, and time-slowing all progressed together. It is difficult to estimate the time period over which these changes occurred, since the sense of time was itself affected. However, there was a feeling of continuous change, rather than a discrete jump or jumps to a new state. Eventually, the sense of time passing stopped entirely. It is difficult to describe this feeling, but perhaps it would be better to say that there was no time, or no sense of time. Only the present moment existed. My elation proceeded to an ecstatic state, the intensity of which I had never even imagined could be possible. The white light around me merged with the reddish light of the sunset to become one all enveloping, intense undifferentiated light field. Perception of other things faded. Again, the changes seemed to be continuous.

At this point, I merged with the light and everything, including myself, became one unified whole. There was no separation between myself and the rest of the universe. In fact, to say that there was a universe, a self, or any "thing" would be misleading—it would be an equally correct description to say that there was "nothing" as to say that there was "everything." To say that subject merged with object might be almost adequate as a description of the entrance into Cosmic Consciousness, but during Cosmic Consciousness there was neither "subject" nor "object." All words or discursive thinking had stopped and there was no sense of an "observer" to comment or to categorize what was "happening." In fact, there were no discrete events to "happen"—just a timeless, unitary state of being.

Cosmic Consciousness is impossible to describe, partly because describing involves words and the state is one in which there were no words. My attempts at description here originated from reflecting on Cosmic Consciousness soon after it had passed and while there was still some "taste" of the event remaining.

Perhaps the most significant element of Cosmic Consciousness was the absolute knowingness that it involves. This knowingness is a deep understanding that occurs without words. I was certain that the universe was one whole and that it was benign and loving at its ground. Bucke's experience was similar. He knew "that the universe is so built and ordered that without any peradventure all things work together for the good of each and all, that the foundation principle of the world is what we call love and that the happiness of everyone is in the long run absolutely certain."

The benign nature and ground of being, with which I was united, was God. However, there is little relation between my experience of God as ground of being and the anthropomorphic God of the Bible. That God is separate from the world and has many human characteristics. "He" demonstrates love, anger and vengeance, makes demands, gives rewards, punishes, forgives, etc. God as experienced in Cosmic Consciousness is the very ground or "beingness" of the universe and has no human characteristics in the usual sense of the word. The universe could no more be separate from God than my body could be separate from its cells. Moreover, the only emotion that I would associate with God is love, but it would be more accurate to say that God is love than God is loving. Again, even characterizing God as love and the ground of being is only a metaphor, but it is the best that I can do to describe an indescribable experience.

The knowingness of Cosmic Consciousness permanently convinced me about the true nature of the universe. However, it did not answer many of the questions that (quite rightly) seem so important to us in our usual state of consciousness. From the perspective of Cosmic Consciousness, questions like, "What is the purpose of life?" or "Is there an afterlife?" are not answered because they are not relevant. That is, during Cosmic Consciousness ontologic questions are fully answered by one's state of being and verbal questions are not to the point.

Eventually, the Cosmic Consciousness faded. The time-changes, light, and mood-elevation passed off. When I was able to think again, the sun had set and I estimate that the event must have lasted about twenty minutes. Immediately following return to usual consciousness, I cried uncontrollably for about a half hour. I cried both for joy and for sadness, because I knew that my life would never be the same.[8]

Afterward, Allan Smith changed the course of his career. Although he says that his experience did not make him "an instant saint or enlightened being," he found that he had less anxiety in general and was able to truly enjoy life. In fact, research has shown that MEs usually leave a permanent mark on the experiencer.

||||

A century ago, Evelyn Underhill, a British writer and a mystic herself, published *Mysticism,* an important guide to the views and practices of Western mystics.[9] In this classic book, she writes that the essence of the mystic life consists of the reshaping of personality. In keeping with this, studies have demonstrated that MEs often lead to profound transformative changes in attitudes and behavior—changes in one's worldview, beliefs and values, relationships, and sense of self.[10] MEs tend also to give the experiencer a sense of purpose and a new meaning to life. People who undergo MEs display lower scores on psychopathology scales and higher psychological well-being than people who never had this kind of experience. In addition, mystic experiencers commonly gain a totally new perspective on their everyday problems, and no longer feel victimized by their daily trials and tribulations.[11]

In 1975, Father Andrew Greeley, a Roman Catholic priest and a sociologist, conducted a survey about the incidence of MEs as a research associate with the National Opinion Research Center (NORC) at the University of Chicago.[12] In this survey, 35 percent of 1,460 respondents answered affirmatively to the question, "Have you ever felt as though you were close to a powerful, spiritual force that seemed to lift you out of yourself?" A few years later, a follow-up study investigated specifically the experiences that gave rise to "Yes" answers in the 1975 survey.[13] This study found that approximately 1 percent of the total sample seemed to have had deep MEs. This finding suggests that each year, there are probably many thousands of people, worldwide, who have such experiences.[14] The question science asks is, Where do they come from?

||||

The ancient Greeks believed that the visions experienced by persons during epileptic seizures were sent by the gods. Because of this, they called epilepsy the "sacred disease." Can MEs be really triggered by abnormal brain activity, as some scientists believe? It is an intriguing question.

Neurologists Jeffrey Saver and John Rabin have suggested that the MEs of some of the greatest religious figures in history—Saint Paul, Joan of Arc, Teresa of Avila, and Thérèse of Lisieux—were probably due to temporal lobe epilepsy (TLE).[15] This neurological disorder is characterized by brief disturbances to the normal electrical activity in the temporal lobes, which are located one on each side of the head, just above the ears. Temporal lobes are involved in audition, vision, memory, and emotion. People having temporal lobe seizures typically report auditory and visual hallucinations, dysphoria, fear, anger, or impressions of depersonalization and unreality. Very rarely, however, do they report spiritual experiences.[16]

University of Chicago neurologist John Hughes, who specializes in the treatment of epilepsy, has conducted a detailed investigation about Saint Joan of Arc (also known as the Maid of Orléans). He concludes that this famous religious figure did not in fact have TLE.[17] He also notes that symptoms of TLE have not much in common with the various aspects of MEs.

National surveys in the United States, Britain, and Australia have revealed that 20 to 49 percent of the individuals interviewed have had spiritual experiences.[18] These experiences bring the experiencer into contact with a larger, nonmaterial reality. Although not as impressive as full-blown introvertive MEs, spiritual experiences can nevertheless be transformative. Here is an example of a spiritual experience that was reported to me by a man named John, following the publication of my book *The Spiritual Brain:*[19]

> *Ten years ago I was a shooting myself up with addictive hard drugs. My mind was gone and I was a hopeless case until this experience happened. I never told anyone before (unbelievers and sceptics abound). One night a bright white Light hit me between the eyes. The room lit up and the Light surged through my whole body. I have not used drugs, alcohol or tobacco ever since. This experience saved my life and gave me great faith that I am not alone. This experience also showed that recovery from chronic needle use of cocaine, heroin and speed is possible.*
>
> *I now consider myself a part of the solution in society, not so much a part of the problem anymore.*
>
> *I am not religious by any stretch but I believe in a higher power that diverted my path. I had no faith until this life saving gift happened.*

Given the relatively common incidence of spiritual experiences in the general population, it is fair to say that most people who have spiritual or mystical experiences are not epileptics. If TLE really produced such experiences, all or most people suffering from this disorder would have them. Evidently, TLE does not play the role that Saver and Rabin have suggested.

▬▬

Michael Persinger is a controversial neuroscientist working at Laurentian University in Sudbury, Ontario, Canada. This self-proclaimed atheist materialist has proposed that small, transient, electrical seizures within the temporal lobes can account for OBEs, visitations by spiritual entities, alien abductions, religious experiences, and mystical visions. Persinger also speculated that it is possible to experimentally induce spiritual experiences by stimulating the temporal lobe with weak electromagnetic currents.[20] This hypothesis led to the creation, in the early 1980s, of the "Koren Helmet" (after its inventor, Stanley Koren, a colleague of Persinger's). This modified snowmobile helmet contains magnetic coils placed over the temporal lobes. The coils are designed to stimulate electromagnetically this part of the brain. Journalists began calling this device the "God Helmet" after it appeared in several TV documentaries.

Since the 1980s, Persinger and his colleagues have conducted a series of studies, using this apparatus, to test the hypothesis that spiritual experiences can be triggered by temporal lobe stimulation. In one of those studies,[21] weak, pulsed electromagnetic fields (not much stronger than the ones generated by a cell phone or a computer monitor) were applied over the temporal area for twenty minutes while participants—psychology students—were wearing opaque goggles in a very quiet room. Another group of students was exposed to a sham field condition—that is, they were not exposed to an electromagnetic field, although all participants were instructed that they might be. Under the influence of the electromagnetic fields, two-thirds of the participants reported a "sensed presence"—the sense that someone else was with them. But 33 percent of the control (sham field) group also reported a sensed presence. Participants attributed the sensed presence to a "spirit guide" or a deceased member of the family.

A few years ago, a research team at Uppsala University in

Sweden, led by psychologist Pehr Granqvist, attempted to replicate the work of Persinger and his colleagues.[22] The Swedish team used Persinger's equipment and consulted Persinger's collaborator Stanley Koren to make certain that conditions for replication were flawless. Granqvist and his co-workers ensured that their experiment was a double-blind by using two experiment conductors who were not told about the goal of the study. The first experimenter interacted with the participants while the second experimenter switched the electromagnetic fields off or on without advising either the first experimenter or the participant. In this way neither the participants nor the experimenters knew who was being exposed to the electromagnetic fields.

Study participants included eighty-nine undergraduate psychology and theology students. They were told that the study sought to examine the influence of weak electromagnetic fields on experiences and feeling states. Participants were not aware that there was a sham-field (control) condition. So the researchers could evaluate the impact of personality traits on the results, participants were asked to fill out a questionnaire after they completed the experiment.

No evidence was found for an effect of the weak electromagnetic fields. Personality traits—in particular, suggestibility—were the best predictor of the outcomes of the study. Of the three participants who reported strong spiritual experiences, two were members of the control group. Of the twenty-two who reported "subtle" experiences, eleven were members of the control group. Those participants all scored high on the suggestibility trait. Granqvist and colleagues attributed their findings to the fact that they had used a double-blind, randomized, controlled procedure. They also argued that Persinger's team's studies were not really double-blind, since the persons conducting the experiments knew what sort of results to expect. The Swedish researchers also concluded that psychological suggestion was the best explanation for the results of the investigations headed by Persinger.

As expected, Persinger contested the findings of the Uppsala research team. He contended that they did not use the helmet properly or for an adequate period of time. Granqvist rejected these arguments by saying that Persinger had agreed to all the details of the stimulation procedures.

The brief descriptions of the experiences reported by the participants in Persinger's studies bear very little resemblance to genuine spiritual experiences. These descriptions do not support Persinger's claim that he and his colleagues have been able to induce authentic spiritual experiences by applying weak electromagnetic fields over the temporal lobes. Such a conclusion should not come as a surprise given that studies involving direct electrical stimulation of the temporal lobes in epileptic patients have failed to produce spiritual experiences. As a matter of fact, direct electrical stimulation of the temporal lobes very rarely elicit any mental response, such as sensations, images, thoughts, and emotional feelings.[23] This does not mean, however, that that the temporal lobe is not implicated in the neural mediation of spiritual experiences.

||||

Radiologist and neuroscientist Andrew Newberg is a pioneer in an emerging field that has been called *neurotheology* by various journalists and authors. The main objective of researchers in neurotheology is to uncover what happens in the brain during religious and spiritual experiences. Nearly a decade ago, Newberg and some colleagues at the University of Pennsylvania scanned Franciscan nuns while they performed a "centering prayer" to open themselves to the presence of God.[24] This prayer involved a mental repetition of a particular phrase. The nuns were scanned with single photon emission computed tomography (SPECT), an imaging technique for mapping blood flow and metabolism after the injection of radioactive substances. Subjectively, the nuns reported that during prayer, they experienced a loss of their usual sense of space. Objectively, the SPECT showed blood flow changes in the superior parietal lobule, an area located in the posterior portion of the parietal lobe. Newberg and his co-workers proposed that the altered sense of space experienced by the nuns during the centering prayer was related to the blood flow changes in this region, which is known to be implicated in the representation of the body within space.

Not long after the publication of this trailblazing work, I undertook a research program concerned with what is happening in people's brains when they experience mystical states of consciousness. For our first study, I recruited nuns who were members of the Carmelite Order. Mystical life occupies a central place in this Roman Catholic organization, which was founded during the twelfth century by a group of hermits living on Mount Carmel in northern Israel. Mystics Saint Teresa of Avila and Saint John of the Cross belonged to this order. Carmelite nuns spend a whole lot of time in silent prayer and contemplation. They are allowed to talk to each other only during two twenty-minute periods after lunch and dinner.

Unsurprisingly, recruiting Carmelite nuns proved to be quite difficult. They had to be reassured that the goal of the study was not to debunk MEs, that is, to demonstrate that these experiences are simply delusions created by a misfiring brain. Overall we found, from convents around the province of Quebec, fifteen Carmelite nuns who agreed to briefly leave their cloistered lives to participate in the research project. All said they had experienced an "intense union with God" at least once. At the time of the study, these fifteen nuns had spent altogether about 210,000 hours in prayer and contemplation.

In the brain imaging lab, the nuns were asked to remove their habits and veils and put on T-shirts and hospital pants. They were given earplugs, and foam cushions were placed around their heads to reduce the loud buzzing and clicking sounds produced by the MRI scanner. The nuns had told us before the onset of the study that "God can't be summoned at will," so they were asked to relive the most intense ME they had felt as a member of the Carmelite Order. This was not too problematic given that when individuals are requested to recall and relive an emotionally charged autobiographical event, they tend to activate the same brain regions and networks as they would activate during the event itself.

Immediately at the end of the experiment, the nuns were asked to evaluate their experiences with the Hood's Mysticism Scale, a questionnaire aimed at measuring reported MEs. The main items from this scale associated with the nuns' experiences were *I have had an experience which I knew to be sacred, I have had an experience in which something greater than myself seemed to absorb me*, and *I have experienced profound joy*. During the qualitative interviews conducted at the end of the experiment, the nuns reported that they had felt the presence of God, his unconditional and infinite love, as well as plenitude and peace. The MEs of the nuns were associated with activation not only in the temporal lobe but also in several other brain regions implicated in perception, positive emotions, body representation within space, and self-consciousness.[25] These findings suggest that there is no single "God spot" or "God module" in the temporal lobes.

Recently, my research team and I conducted another fMRI study, this time in NDErs who claim to have remained in contact with the "light" encountered in their NDE.[26] Fifteen NDErs living in the Montreal area were recruited for this study. All these people had nearly died or had experienced clinical death and were revived. Again, earplugs and foam cushions were used to diminish the noise generated by the MRI scanner. The NDErs were scanned while, in a meditative state, they attempted to "reconnect" with the light.

Many of the participants were crying when they came out of the scanner. One of them, Jorge Medina, a man who had had an NDE following a fire that left 90 percent of his body with third-degree burns, even reported the impression of leaving his body during scanning. As in our study of the Carmelite nuns, many brain areas were activated, including the temporal lobe. However, there was only a small overlap between the regions activated in the nuns and those lighting up in the NDErs. This indicates that different spiritual experiences are associated with distinct brain areas and networks.

In brain imaging studies, correlations—statistical measures of association—are established between the cerebral regions activated and changes in the mental activity of the individuals being scanned. Such correlations do not allow neuroscientists to make causal conclusions. In the case of our fMRI studies, this means that we cannot affirm that the activation in the various regions identified was actually the cause of the spiritual experiences reported by the nuns and the NDErs.

Obviously, these brain imaging studies cannot prove or disprove the existence of a "Higher Power." Certain researchers and journalists have argued that the fact that spiritual experiences are associated with neural correlates suggests that such experiences are merely delusions—"nothing but" brain activity. This is a mistaken view, equivalent to assuming that the painting you are contemplating is an illusion because it is associated with identifiable brain activity in the visual portion of your brain.

〓

Throughout history, many religious rituals aimed at occasioning mystical states of consciousness have involved the use of plant-derived substances known as psychedelics (psychedelic means "mind-manifesting"). These psychoactive substances have played a major role in spiritual practices of indigenous societies of the Amazon, Central America, and Siberia.[27] The morning glory vine, whose active compounds are psilocybin and psilocin, were used in religious ceremonies by the Aztecs, and are still used today by tribes in Mexico to communicate with the spirit world. The amateur ethnobotanist Gordon Wasson suggested, four decades ago, that the *amanita muscaria* mushroom was the sacred Soma plant celebrated in the Rig Veda, an ancient collection of Vedic Sanskrit hymns.[28] The peyote cactus, whose main active principle is mescaline, was used by the Aztecs as far as 300 B.C. Peyote is still utilized for spiritual purposes today by members of the Native American Church.

Based on his own experiences with mescaline, the famed author Aldous Huxley proposed that all humans are potentially "mind at large"; that is, they are part of the Ground of Being.[29] He argued that the brain and the nervous system act as a filter (or reducing valve) to render biological survival possible, and surmised that psychedelic agents are able to temporarily impair or disable this filter function, allowing the experience of mind at large.

Some of the first studies on psychedelics were conducted at Harvard University in the 1950s and 1960s. Huston Smith participated in a few of these studies. This MIT religious studies scholar became famous after the publication, in 1958, of *The Religions of Man*—the most popular textbook ever written about the world's religions.[30] Keenly aware that consciousness-altering substances have been associated with religion throughout history and across the world, Smith wanted to verify whether certain psychedelics can truly facilitate MEs.

On New Year's Day 1961, Huston Smith took an average dose of mescaline in the presence of Harvard psychologist Timothy Leary. Later, he wrote:

I was experiencing the metaphysical theory known as emanationism, in which, beginning with the clear, unbroken and infinite light of God or the void, the light then breaks into forms and decreases in intensity as it diffuses through descending degrees of reality. My friends in the study were functioning in an intelligible wave band, but one which was far more restricted, cramped and wooden than the bands I was now privileged to experience. Bergson's notion of the brain as a reducing valve seemed accurate.

Along with "psychological prism," another phrase came to me: "empirical metaphysics." The emanation theory and elaborately delineated layers of Indian cosmology and psychology had hitherto been concepts and inferences. Now they were objects of direct, immediate perception. I saw that theories such as these were required by the experience I was having. I found myself amused, thinking how duped historians of philosophy had been in crediting those who formulated such world views with being speculative geniuses. Had they had experiences such as mine they need have been no more than hack reporters. But beyond accounting for the origin of these philosophies, my experience supported their truth. As in Plato's myth of the cave, what I was now seeing struck me with the force of the sun in comparison with which normal experience was flickering shadows on the wall.

How could these layers upon layers, these worlds within worlds, these paradoxes in which I could be both myself and my world, and an episode could be both instantaneous and eternal—how could such things be put into words? I realized how utterly impossible it would be for me to describe them . . .[31]

Subsequently, Huston Smith was recruited for what became known as the "Good Friday Experiment." This experiment was led by Walter Pahnke, a physician then completing a doctoral degree in religious studies at Harvard. Pahnke used a double-blind approach in which one group of participants (mostly graduate students) received psilocybin—the active component in so-called magic mushrooms—while the other group was given niacin (as a placebo), a component of the vitamin B complex found in meat, dairy products, and wheat germ. Participants in both groups did not know who had received the psychedelic. The experiment took place in Marsh Chapel at Boston University before the Good Friday service of 1962. Smith found himself in the group that was given psilocybin. Nearly all members of the psilocybin group, and only the psilocybin group, reported experiences that could not be distinguished from MEs. In line with Huxley's view, Huston Smith's own experimentation with psychedelics convinced him that these substances can occasion genuine MEs.[32]

Unfortunately, psychedelic substances became taboo among federal regulators after Timothy Leary promoted the use of LSD with the slogan "Turn on, tune in, drop out." The culture war that followed led the federal government to shut down, in the early 1970s, most human-based research into psychedelic agents.

IIII

In the 1990s, federal regulators resumed granting approval for controlled human studies of psychedelic effects. Researchers at universities such as Harvard, Johns Hopkins, the University of Arizona, and the University of California–Los Angeles (UCLA) are now carrying out human research with psychedelics. This work is supported mainly by nonprofit groups like the Heffter Research Institute and the Multidisciplinary Association for Psychedelic Studies (MAPS).

Neuropharmacologist Roland Griffiths and his Johns Hopkins University colleagues recently attempted to replicate Pahnke's experiment. They recruited thirty-six people who had no serious physical or emotional problems. None had had any previous experience with psychedelics. Almost all participants—mostly middle-aged college graduates—engaged in spiritual activities such as prayer and attending religious services.

The experiment was double-blind—neither the participants nor the monitors knew whether the participants were receiving psilocybin or the control drug. Ritalin, a stimulant drug that paradoxically calms hyperactive children, was selected as the control. On one of two occasions, the experimenters gave the volunteers either psilocybin or Ritalin, swapping the drugs on the second occasion. Participants were encouraged to close their eyes and direct their attention inward. They were monitored for eight hours after the drugs were administered, to ensure all was well. None of the participants reported any serious negative effects during the experiment. Immediately after and again two months following sessions, they were asked to complete questionnaires assessing drug effects and ME. The Pahnke–Richards Mystical Experience Questionnaire was used to assess the experiences of the participants. This questionnaire, which was designed specifically to evaluate MEs, is based on the descriptive work done by Walter Stace.[33]

Two-thirds of the participants who received psilocybin rated it as either the best experience of their lives or within the top five. These participants described having the sense that the boundaries between the self and others had disappeared and belonging to some larger state of consciousness. They also described a sense of the unity of all things, and feelings of peace and intense happiness. Two months after the study, 79 percent of them reported moderately or greatly enhanced well-being or satisfaction. Their positive changes in general feelings, attitudes, and behavior were confirmed independently by assessments made by family and friends.

This landmark study was published in the *Journal of Psycho-pharmacology* in 2006.[34] In a follow-up survey, taken fourteen months after the experiment, most of these participants once again expressed more well-being and rated the experience as one of the five most meaningful events of their lives.[35] Given the striking resemblance between participants' reports and accounts of MEs by mystics, Griffiths believes that the human brain is probably wired to undergo these "unitive" experiences.[36]

I agree with Henri Bergson and Aldous Huxley that our usual states of brain activity produce a filter function that generally renders us unaware of the Ground of Being. The neuroscientific studies presented throughout this chapter suggest that alterations in electrical and chemical activities in the brain are necessary for MEs to take place. As we have seen, such alterations can be occasioned by a variety of means and it is thus likely that different patterns of brain activity can mediate MEs. The case of Pam Reynolds, presented in the previous chapter, and other cases of NDEs induced by cardiac arrest also indicate that deep MEs can occur when the brain is not functioning. This is understandable given that the brain's filter function is then inoperative.

||||

MEs, especially of the introvertive kind, teach us a number of crucial things that fly in the face of the scientific materialist worldview. These experiences tell us that contrary to appearances, we are not encapsulated within our brains and bodies and separate from each other but, rather, "organically" connected with all others and with everything in the entire universe. Psychiatrist Stanislav Grof, one of the founders of transpersonal psychology, notes that the experience of the farther reaches of the human psyche reveals that mind and consciousness are fundamental features of all existence, and they are closely intertwined with the physical world.[37] MEs also suggest that just as the drop of the ocean is the ocean itself, we are part of an infinite cosmic intelligence that pervades everything in the whole universe, beyond time and space.

In mystical states of consciousness, we can transcend the illusory boundaries of our "little self" and experientially connect—within the innermost level of the psyche—with this intelligence, the Ground of Being, which is our "true self." During such states, *the drop realizes she is the ocean*. In the conclusion of this book, I present an emerging model of reality that goes far beyond the materialist worldview to encompass the drop and the ocean as part of the same boundless world, making sense of what dogmatic materialists call nonsense.

A Great Shift in Consciousness

The Universe begins to look more like a great
thought than like a great machine.

—PHYSICIST AND ASTRONOMER JAMES JEANS[1]

In the quantum universe—in which we increasingly live—there is no mind-brain problem because there is no radical separation between the mental world and the physical world. The new paradigm is here; we just need to open our eyes. It is my fervent hope that *Brain Wars* will be an important part of that process.

The wealth of scientific studies you have read about in the chapters of this book indicate that our thoughts, beliefs, and emotions can greatly influence what is happening in our brains and bodies and play a key role in our health and well-being:

Norman Cousins was one of a number of people who have demonstrated that our beliefs and expectations about medical treatments can stimulate our self-healing capacity, even in diseases as severe as cancer and Parkinson's disease. And recent research suggests that our thoughts and emotions can even affect how our bodies can turn certain genes on or off.

Jake's success in controlling his seizures and behaviors is one of many examples of how we can use neurofeedback to deliberately change brain processes that are normally not under voluntary control and improve our mental functions. As you have seen, other studies show that we can intentionally train our minds—through meditative practices—to enhance the activity of brain areas implicated in emotional well-being, compassion, and attention. Meditative practices can even alter the physical structure of the brain.

Our minds can be extremely powerful—far more powerful than we thought only a few decades ago.

The effects of the mind and mental abilities are not limited to the confines of the body. For instance, psi studies show that we can sometimes receive meaningful information without the use of ordinary senses and in ways that transcend the habitual space and time constraints. Still other psi research demonstrates that we can intentionally influence—at a distance—not only random number generators but living organisms, including human beings.

NDE studies show that people like Pam Reynolds can have veridical perceptions—corroborated by independent witnesses—during OBEs triggered by a cardiac arrest. These perceptions concern events that occur *while the heart is not functioning*. We know that the activity of the brain ceases within a few seconds following a cardiac arrest. Given this, the findings of NDE research strongly challenge the idea that mind is "only" a product of brain activity, giving rather more credence to the view that mind may be dependent on the brain "much as a radio transmission is dependent upon a receiver and broadcast unit."[2] Additionally, the mystical (or transcendental) component of NDEs occurring during a cardiac arrest supports the idea that the brain usually acts as a filter that prevents the perception of what could be dubbed *other realms of reality*. This aspect of NDEs also corroborates the idea that we are more than our physical bodies.

Mental activity is not the same as brain activity,[3] and we are not "meat puppets," totally controlled by our brains, our genes, and our environments. Indeed, our minds and our consciousness can significantly affect events occurring in the brain and body, and outside the body. We do have these immensely important capacities, and it is time for science to begin taking them seriously. But for this to happen, science—and all of us—must change the lens through which we view reality.

Fortunately, the scientific enterprise (as a method, not as materialist ideology) allows for all of these possibilities, and infinitely more. Materialist science, based on the classical Newtonian physics, took science out of the Dark Ages, showing us a world no one had ever seen before. Now there is another heretofore invisible world for us to see, one that the dogmas of materialist science obscure but that is brought into focus by the discoveries of quantum physics.

||||

Toward the end of the nineteenth century, it became obvious that classical physics was limited; it was just not able to explain certain phenomena at the atomic level. The acknowledgment of these limitations led to the development of a revolutionary new branch of physics called quantum mechanics (QM), which smashed the scientific materialist worldview. In the words of theoretical physicist Amit Goswami, QM is "a new paradigm of science based on the primacy of consciousness. . . . The new paradigm resolves many paradoxes of the old paradigm and explains much anomalous data."[4]

The work of QM has effectively *dematerialized* the classical universe by showing that it is not made of minuscule billiard balls, as drawings of atoms and molecules would lead us to believe. QM has shown that atoms and subatomic particles are not really *objects*—they do not exist with certainty at definite spatial locations and definite times. Rather, they show "tendencies to exist," forming a world of potentialities within the quantum domain.[5] Werner Heisenberg, winner of the 1932 Nobel Prize in Physics, explained, "The atoms or elementary particles themselves are not real, they form a world of potentialities or possibilities rather than one of things or facts."[6]

The quantum world appears different from the physical world, but some elements may sound familiar. For example, a central feature of QM is called the *observer effect:* particles being observed and the observer—the physicist and the method used for observation—are linked, and the results of the observation are influenced by the observer's conscious intent. This effect implies that the consciousness of the observer is vital to the existence of the physical events being observed. In other words, QM acknowledges that the physical world cannot be fully understood without making reference to mind and consciousness.[7]

In QM, the physical world is thus no longer viewed as the primary or sole component of reality. Most contemporary physicists agree with Wolfgang Pauli—one of the founders of QM—that the physical and the psychological, *physis* and *psyche,* should be recognized and embraced as distinct but complementary aspects of one reality.[8] Regarding this issue, the mathematician and physicist John von Neumann raised the possibility that mind and consciousness constitute not an emergent property but rather a fundamental component of the universe. Regardless of whether it is the case, QM teaches us that we must consider mind and consciousness if we are to reach a more adequate conception of nature and reality.

Nonlocality (or nonseparability)—which Albert Einstein memorably referred to as "spooky actions at a distance"—is another remarkable discovery of QM. This concept is based on entanglement, the instantaneous connections that persist between particles (such as photons, electrons) that interacted physically and then become separated. These connections remain even if the particles are separated by enormous distances (for instance, billions of light-years). This counterintuitive aspect of nature has been demonstrated experimentally in a number of labs since the beginning of the 1970s. Nonlocality and entanglement suggest that the universe constitutes an undivided whole.[9]

||||

Naturally, contemporary materialists strongly disagree with the conclusion that scientific materialism has failed and is currently breaking down. They argue that sooner or later, neuroscience will be able to completely explain mind and consciousness.[10] These materialists do not seem to realize that future technological development will only allow neuroscientists to measure more refined correlates of mental activity.

Belief in the materialist worldview compels certain scientists and philosophers to neglect the subjective dimension of human experience and downplay the importance of mind and consciousness. In so doing, they create a severely distorted and impoverished understanding of human beings and reality.

Materialist scientists and philosophers are also led to consider certain phenomena such as psi, NDE, and mystical experiences (MEs) as anomalous. These phenomena are anomalous only to the extent that we cling to the false assumptions of scientific materialism. Seen and understood through the lens of QM, most of these phenomena do not appear anomalous at all. So-called paranormal events are, in effect, perfectly normal.

IIII

Physicists were forced to abandon the assumptions of classical physics and the scientific materialist worldview nearly a century ago, but the battles of the brain wars are still being fought by many neuroscientists. The time has come for my colleagues to embrace the many possibilities of the universe opened by the new physics and free their minds from the shackles and blinders of the scientific materialist credo.

The expanded model of reality you have read about in this book offers infinite and exciting concepts for science—and you—to freely explore:[11]

- This model acknowledges *all* the empirical evidences related to mind and consciousness, not only those that appear at first sight to be compatible with materialist theories.

- It includes the mental and the physical, the subjective and the objective, the first-person perspective, and the third-person perspective.

- It assumes that mind and consciousness are a prerequisite for reality because they allow us to perceive and experience the world. Stated otherwise, they represent an aspect of reality as fundamental as the physical world.

- It assumes that mind and the physical world are continually interacting because they are not really separated—*they only appear to be separated.* This means that there is a deep interconnectedness between the mental world and the physical world, which both arise out of the same source. This basic interconnectedness renders the mind capable of influencing various phenomena and events belonging to the physical world. Information may act as a bridge between these complementary aspects of reality. Some physicists go as far as to suggest that the whole of reality can be seen as a pattern of information.

- It assumes that mind and consciousness are not produced by the brain. This idea suggests that mental functions and personality can survive physical death. In other respects, MEs indicate that we are not encapsulated within our brains and bodies but, rather, connected—within the deepest levels of the psyche—with everything in the universe, as well as with the underlying source giving birth to both mind and matter. In this way, MEs represent a direct, intuitive apprehension of the undivided wholeness.

The scientific evidence you have read about in this book makes two things clear: scientific materialism is just plain false, and we humans are not powerless, biochemical machines. Together with exciting possibilities of the quantum universe, this evidence tells us that it is time to enlarge our concept of the natural world to reintegrate mind and consciousness.

This emerging scientific model of reality—this new paradigm of what is possible—has far-reaching implications. Perhaps most important, it fundamentally alters the vision we have of ourselves, giving us back our dignity and power, as humans and as scientists.

We are no longer at the mercy of Big Pharma: in many instances we can willfully choose to positively influence our health and mental functioning by being aware of our thoughts and emotions, and by training our brains.

Scientists, free of the materialist box, are now invited to embark on research into the whole gamut of psi phenomena, expanded and altered consciousness, and spiritual experiences.

Last but not least, the new paradigm fosters positive values such as compassion, respect, and peace. By emphasizing a deep connection between ourselves and nature at large, it also promotes environmental awareness and the preservation of our biosphere.

When mind and consciousness are recognized as one, we are again connected to ourselves, to each other, to our planet, and to the universe.

A great shift in consciousness has begun, bringing with it a profound transformation of our world.[12]

ACKNOWLEDGMENTS

I am very grateful to my literary agent, Loretta Barrett, for her assistance, efficiency, and professionalism; and to Gideon Weil, executive editor at HarperOne, for his excellent editorial suggestions and his contagious enthusiasm. I also wish to thank Naomi Lucks for her thorough revision of the manuscript. Last, I thank my wife, Johanne, for her enduring love, support, and patience.

NOTES

Introduction : A "Computer Made of Meat"?

1. Arthur Koestler, *Janus: A Summing Up* (New York: Random House; 1978), 229. Cited in Larry Dossey, "Is the Universe Merely a Statistical Accident?," *HuffPost Living,* June 23, 2010, http://www.huffingtonpost.com/dr-larry-dossey/spiritual-living-is-the-u_b_621261.html.

2. Francis Crick, *The Astonishing Hypothesis: The Scientific Search for the Soul* (New York: Simon & Schuster, Touchstone, 1995), 3.

3. Dossey, "Is the Universe Merely a Statistical Accident?"

4. M. R. Bennett, "Development of the Concept of Mind," *Royal Australian and New Zealand College of Psychiatrists* 41 (2007): 943–56.

5. Bennett, "Development of the Concept of Mind."

6. Richard Tarnas, *The Passion of the Western Mind: Understanding the Ideas That Have Shaped Our World View* (New York: Ballantine Books, 1991).

7. E. M. Sternberg and P. W. Gold, *The Mind-Body Interaction in Disease*, http://being.publicradio.org/programs/stress/mindbodyessay.shtml.

8. Tarnas, *Passion of the Western Mind.*

9. Thomas Huxley, "On the Hypothesis That Animals Are Automata, and Its History," *Fortnightly Review* 16 (1874): 555–80.

10. Ferdinand Schiller, *Riddles of the Sphinx* (London: Swan Sonnenschein, 1891).

11. William James, "Human Immortality: Two Supposed Objections to the Doctrine," in *William James on Psychical Research,* ed. G. Murphy and R. O. Ballou (New York: Viking; 1898), 279–308.

12. James, "Human Immortality."

13. Wilder Penfield, *Control of the Mind*. Symposium held at the Univ. of California Medical Center–San Francisco, 1961, quoted in Arthur Koestler, *Ghost in the Machine* (London: Hutchinson, 1967), 203–4.

14. Wilder Penfield, *The Mysteries of the Mind* (Princeton, NJ: Princeton Univ. Press, 1975).

15. John C. Eccles, *Evolution of the Brain: Creation of the Self* (London: Routledge, 1989), 241.

16. John C. Eccles, *How the Self Controls Its Brain* (Berlin/New York: Springer, 1994); F. Beck and J. C. Eccles, "Quantum Aspects of Brain

Activity and the Role of Consciousness," *Proceedings of the National Academy of Sciences of the United States of America* 89 (1992): 11357–61.

Eccles proposed that dendrites—branched extensions of a neuron that conducts impulses from adjacent neurons inward toward the cell body—are the basic receptive units of the cerebral cortex (the mantle of gray matter covering the surface of the cerebral hemispheres). Bundles of dendrites (called by Eccles *dendrons*) are linked with mental units (or *psychons*) that represent a specific experience.

In mental intentions and willed thoughts, he said, psychons act on dendrons and momentarily increase the probability of release of chemical messengers from presynaptic vesicles (the small, membrane-bound vesicles at a synaptic junction of neurons that contain the chemical messengers). This process does not violate the conservation law because the extremely small size of the synaptic "microsite" that emits the chemical messengers is within the range of quantum mechanics.

17. H. Feigl, "The Mental and the Physical," in *Minnesota Studies in the Philosophy of Science: Concepts, Theories, and the Mind-Body Problem*, Vol. 2, ed. H. Feigl, M. Scriven, and G. Maxwell (Minneapolis: Univ. of Minnesota Press, 1958).

18. P. M. Churchland, *A Neurocomputational Perspective: The Nature of Mind and the Structure of Science* (Cambridge, MA: MIT Press, 1992).

19. R. W. Sperry, "Mind-Brain Interaction: Mentalism, Yes; Dualism, No," *Neuroscience* 5 (1980): 195–206.

20. See, for instance, N. Murphy and W. S. Brown, *Did My Neurons Make Me Do It? Philosophical and Neurobiological Perspectives on Moral Responsibility and Free Will* (New York: Oxford Univ. Press, 2007).

21. D. J. Chalmers, "Facing Up to the Problem of Consciousness," *Journal of Consciousness Studies* 2 (1995): 200–219.

22. But the scientific world is far from a consensus about the nature of the relationship between mind and brain. There are many other theories that attempt to account for the relationship between mind and brain, such as functionalism, biological naturalism, neutral monism, and panpsychism. I do not present all the theories in this chapter since this book is written mainly for laypeople. Readers who want to learn more about these theories may consult scholarly works such as *The Cambridge Handbook of Consciousness* and *The Blackwell Companion to Consciousness*.

23. About the work and views of Frederic Myers, I invite readers to read the superb book, E. F. Kelly and E. W. Kelly, eds., *Irreducible Mind: Toward a Psychology for the Twenty-First Century* (Lanham, MD: Rowman & Littlefield, 2007).

24. In science, not only do observations guide theories, but theories also influence the *collection of observations* through experiments.

Chapter 1: The Power of Belief to Cure or Kill

1. Anne Harrington, *The Placebo Effect: An Interdisciplinary Exploration* (Cambridge, MA: Harvard Univ. Press, 1997), 1.

2. B. Klopfer, "Psychological Variables in Human Cancer," *Journal of Projective Techniques and Person Assessment* 21 (1957): 331–34.

3. M. Brooks, "Anomalies: 13 Things That Don't Make Sense," *New Scientist,* March 19–25, 2005.

4. A. Hróbjartsson and P. C. Götzsche, "Is the Placebo Powerless? An Analysis of Clinical Trials Comparing Placebo with No Treatment," *New England Journal of Medicine* 344 (2001): 1594–1602.

5. In 2007 I wrote an academic article that reviewed several studies of placebo effects and argued that these studies clearly show that we can greatly influence our brains and bodies with just our beliefs. Chapter 1 constitutes a vulgarized sequel to this article. M. Beauregard, "Mind Does Really Matter: Evidence from Neuroimaging Studies of Emotional Self-Regulation, Psychotherapy, and Placebo Effect," *Progress in Neurobiology* 81 (2007): 218–36.

6. A. K. Shapiro and L. A. Morris, "The Placebo Effect in Medicine and Psychological Therapies," in *Handbook of Psychotherapy and Behavior Change: An Empirical Analysis*, 2nd ed., ed. S. L. Garfield and A. Bergin (New York: Wiley, 1978), 369–410.

7. T. J. Kaptchuk, C. E. Kerr, and A. Zanger, "Placebo Controls, Exorcisms, and the Devil," *Lancet* 374 (2009): 1234–35.

8. H. K. Beecher, "The Powerful Placebo," *Journal of the American Medical Association* 159 (1955): 1602–6.

9. G. S. Kienle and H. Kienle, "The Powerful Placebo Effect: Fact or Fiction?," *Journal of Clinical Epidemiology* 50 (1997): 1311–18.

10. F. Benedetti, *Placebo Effects: Understanding the Mechanisms in Health and Disease* (New York: Oxford Univ. Press), 2009.

11. R. Ornstein and D. Sobel, *The Healing Brain: Breakthrough Discoveries About How the Brain Keeps Us Healthy* (New York: Simon & Schuster, 1987).

12. D. Evans, *Placebo: The Belief Effect* (London: HarperCollins, 2004).

13. D. E. Moerman, "Cultural Variations in the Placebo Effect: Ulcers, Anxiety, and Blood Pressure," *Medical Anthropology Quarterly* 14 (2000): 51–72.

14. A. J. de Crean et al., "Effect of Colour of Drugs: Systemic Review of Perceived Effect of Drugs and of Their Effectiveness," *British Medical Journal* 313 (1996): 1624–26.

15. I. Hashish et al., "Reduction of Postoperative Pain and Swelling by Ultrasound Treatment: A Placebo Effect," *Pain* 33 (1988): 303–11.

16. J. D. Levine, N. C. Gordon, and H. L. Fields, "The Mechanism of Placebo Analgesia," *Lancet* 312 (1978): 654–57.

17. J. K. Zubieta et al., "Placebo Effects Mediated by Endogenous Opioid Neurotransmission and ì-opioid Receptors," *Journal of Neuroscience* 25 (2005): 7754–62.

18. R. de la Fuente-Fernandez et al., "Expectation and Dopamine Release: Mechanism of the Placebo Effect in Parkinson's Disease," *Science* 293 (2001): 1164–66.

19. A. J. de Craen et al., "Placebo Effect in the Treatment of Duodenal Ulcer," *British Journal of Clinical Pharmacology* 48 (1999): 853–60.

20. K. Schapira et al., "Study on the Effects of Tablet Colour in the Treatment of Anxiety States," *British Medical Journal* 2 (1970): 446–49.

21. I. Kirsch and G. Sapirstein, "Listening to Prozac but Hearing Placebo: A Meta-analysis of Antidepressant Medication," *Prevention & Treatment* 1 (1998): 2a. Online document at: http://journals.apa.org/prevention.

22. L. A. Cobb et al., "An Evaluation of Internal-Mammary-Artery Ligation by a Double-Blind Technique," *New England Journal of Medicine* 260 (1959): 1115–18.

23. J. B. Moseley Jr. et al., "Arthroscopic Treatment of Osteoarthritis of the Knee: A Prospective, Randomized, Placebo-Controlled Trial: Results of a Pilot Study," *American Journal of Sports Medicine* 24 (1996): 28–34.

24. J. B. Moseley et al., "A Controlled Trial of Arthroscopic Surgery for Osteoarthritis of the Knee," *New England Journal of Medicine* 347 (2002): 81–88.

25. C. McRae et al., "Effects of Perceived Treatment on Quality of Life and Medical Outcomes in a Double-Blind Placebo Surgery Trial," *Archives of General Psychiatry* 61 (2004): 412–20.

26. McRae et al., "Effects of Perceived Treatment on Quality of Life," 418–19.

27. E. D. Eaker, J. Pinsky, and W. P. Castelli, "Myocardial Infarction and Coronary Death Among Women: Psychosocial Predictors from a Twenty-Year Follow-Up of Women in the Framingham Study," *American Journal of Epidemiology* 135 (1992): 854–64.

28. W. P. Kennedy, "The Nocebo Reaction," *Medical World* 95 (1961): 203–5.

29. R. R. Reeves et al., "Nocebo Effects with Antidepressant Clinical Drug Trial Placebos," *General Hospital Psychiatry* 29 (2007): 275–77.

30. Reeves et al., "Nocebo Effects."

31. A. J. Barsky and J. F. Borus, "Functional Somatic Syndromes," *Annals of Internal Medicine* 130 (1999): 910–21.

32. M. G. Myers, J. A. Cairns, and J. Singer, "The Consent Form as a

Possible Cause of Side Effects," *Clinical Pharmacology and Therapeutics* 42 (1987): 250–53.

33. M. A. Flaten and T. D. Blumenthal, "Caffeine-Associated Stimuli Elicit Conditioned Response: An Experimental Model of the Placebo Effect," *Psychopharmacology* 145 (1999): 105–12.

34. M. Shepherd, "The Placebo: From Specificity to the Non-Specific and Back," *Psychological Medicine* 23 (1993): 569–78.

35. R. A. Preston et al., "Placebo-Associated Blood Pressure Response and Adverse Effects in the Treatment of Hypertension," *Archives of Internal Medicine* 160 (2000): 1449–54.

36. J. A. Turner et al., "The Importance of Placebo Effects in Pain Treatment and Research," *Journal of the American Medical Association* 271 (1994): 1609–14.

37. P. Enck, F. Benedetti, and M. Schedlowski, "New Insights into the Placebo and Nocebo Responses," *Neuron* 59 (2008): 195–206.

38. T. Koyama et al., "The Subjective Experience of Pain: Where Expectations Become Reality," *Proceedings of the National Academy of Sciences of the United States of America* 102 (2005): 12950–55.

39. H. Benson, "The Nocebo Effect: History and Physiology," *Preventive Medicine* 26 (1997): 612–15.

40. C. K. Meador, "Hex Death: Voodoo Magic or Persuasion?," *Southern Medical Journal* 85 (1992): 244–47.

41. H. Basedow, *The Australian Aborigenal* (Adelaide, Australia: Preece, 1925). Quoted in W. B. Cannon, "Voodoo Death," *American Anthropology* 44 (1942): 169–81.

42. Meador, "Hex Death."

43. H. Pilcher, "The Science of Voodoo: When Mind Attacks Body," *New Scientist,* May 13, 2009. http://www.newscientist.com/article/mg20227081.100-the-science-of-voodoo-when-mind-attacks-body.html.

44. Meador, "Hex Death."

45. S. Silberman, "Placebos Are Getting More Effective. Drugmakers Are Desperate to Know Why," *Wired Magazine,* September 17, 2009. http://www.wired.com/medtech/drugs/magazine/1709/ff_placebo_effect?currentPage=all.

46. Silberman, "Placebos Are Getting More Effective."

47. Silberman, "Placebos Are Getting More Effective."

48. Silberman, "Placebos Are Getting More Effective."

49. L. Colloca and F. G. Miller, "How Placebo Responses Are Formed: A Learning Perspective," *Philosophical Transactions of the Royal Society B: Biological Sciences* 366 (2011): 1859–69.

Chapter 2: Brain Control

1. J. Robbins, *A Symphony in the Brain: The Evolution of the New Brain Wave Biofeedback,* rev. ed. (New York: Grove Press, 2008), xv.
2. Robbins, *Symphony in the Brain.*
3. Robbins, *Symphony in the Brain.*
4. http://www.aapb.org/.
5. Robbins, *Symphony in the Brain.*
6. Robbins, *Symphony in the Brain.*
7. M. B. Sterman, L. R. Macdonald, and R. K. Stone, "Biofeedback Training of the Sensorimotor Electroencephalogram Rhythm in Man: Effects on Epilepsy," *Epilepsia* 15 (1974): 395–416.
8. M. B. Sterman and L. R. Macdonald, "Effects of Central Cortical EEG Feedback Training on Incidence of Poorly Controlled Seizures," *Epilepsia* 19 (1978): 207–22.
9. M. B. Sterman, "Biofeedback in the Treatment of Epilepsy," *Cleveland Clinic Journal of Medicine* 77 (2010): S60–S67.
10. Russell A. Barkley, *Attention Deficit Hyperactivity Disorder: A Handbook for Diagnosis and Treatment* (New York: Guilford, 1996).
11. S. Mannuzza et al., "Educational and Occupational Outcome of Hyperactive Boys Grown Up," *Journal of the American Academy of Child and Adolescent Psychiatry* 36 (1997): 1222–27.
12. M. Arns et al., "Efficacy of Neurofeedback Treatment in A.D.H.D.: The Effects on Inattention, Impulsivity and Hyperactivity: A Meta-analysis," *Clinical Electroencephalography and Neuroscience* 40 (2009): 180–89.
13. Robbins, *Symphony in the Brain.*
14. J. Lévesque, M. Beauregard, and B. Mensour, "Effect of Neurofeedback Training on the Neural Substrates of Selective Attention in Children with Attention-Deficit/Hyperactivity Disorder: A Functional Magnetic Resonance Imaging Study," *Neuroscience Letters* 394 (2006): 216–21; M. Beauregard and J. Lévesque, "Functional Magnetic Resonance Imaging Investigation of the Effects of Neurofeedback Training on the Neural Bases of Selective Attention and Response Inhibition in Children with Attention-Deficit/Hyperactivity Disorder," *Applied Psychophysiology Biofeedback* 31 (2006): 3–20.
15. Lévesque et al., "Effect of Neurofeedback Training on the Neural Substrates of Selective Attention in Children with Attention-Deficit/Hyperactivity Disorder"; Beauregard and Lévesque, "Functional Magnetic Resonance Imaging Investigation of the Effects of Neurofeedback Training on the Neural Bases of Selective Attention and Response Inhibition in Children with Attention-Deficit/Hyperactivity Disorder."

16. Slow waves produced in this area of the brain do not negatively affect cognitive performance.

17. E. G. Peniston and P. J. Kulkosky, "Alpha-Theta Brainwave Training and Beta-Endorphin Levels in Alcoholics," *Alcoholism: Clinical and Experimental Research* 13 (1989): 271–79.

18. E. G. Peniston and P. J. Kulkosky, "Alpha-Theta Brain Wave Neurofeedback for Vietnam Veterans with Combat Related Post Traumatic Stress Disorder," *Medicine and Psychotherapy* 4 (1991): 1–14.

19. R. C. deCharms, "Applications of Real-Time fMRI," *Nature Reviews Neuroscience* 9 (2008): 720–29.

20. R. C. deCharms et al., "Learned Regulation of Spatially Localized Brain Activation Using Real-Time fMRI," *Neuroimage* 21 (2004): 436–43.

21. R. C. deCharms et al., "Control over Brain Activation and Pain Learned by Using Real-Time Functional MRI," *Proceedings of the National Academy of Sciences of the United States of America* 102(2005): 18626–31.

22. J. P. Hamilton et al., "Modulation of Subgenual Anterior Cingulate Cortex Activity with Real-Time Neurofeedback," *Human Brain Mapping* 32 (2011): 22–31.

23. http://www.nature.com/nature/focus/brain/experiments/.

24. A neuroprosthesis is a device that can substitute a motor or sensory modality that might have been impaired as a result of disease or an injury. A cochlear implant represents an example of neuroprosthesis.

25. J. J. Daly and J. R. Wolpaw, "Brain-Computer Interfaces in Neurological Rehabilitation," *Lancet Neurology* 7 (2008): 1032–43.

26. N. Birbaumer et al., "A Spelling Device for the Paralysed," *Nature* 398 (1999): 297–98.

27. http://www.innovationwatcharchive.com/choiceisyours/choiceisyours.2006.11.30.htm.

28. D. Graham-Rowe, "Dialing with Your Thoughts," April 12, 2011, http://www.technologyreview.com/communications/37357/?p1=A1&a=f.

29. H. Yomogida, "NeuroSky Demos Brain-Controlled Mobile Phone Applications," September 11, 2008, http://www.technologyreview.com/communications/37357/?p1=A1&a=f.

30. M. Snider, "New Toy Trains to Use 'The Force,'" http://www.mindpowernews.com/LearnTheForce.htm.

31. Other conditions treated with neurofeedback include major depression and the Asperger syndrome.

32. http://www.futurehealth.org/populum/page.php?f=Electroencephalography-and-by-Vietta-Sue-Wilson–100208–348.html.

33. http://braintrackers.com/library/76-italian-soccer-training-a-biofeedback.

34. http://www.nba.com/clippers/news/kaman_espnotl_080115.html.

35. http://neurofeedback-singapore.com/news-update.html.

Chapter 3: Train Your Mind, Transform Your Brain

1. The Dalai Lama citation is from S. Begley, *Train Your Mind, Change Your Brain: How a New Science Reveals Our Extraordinary Potential to Transform Ourselves* (New York: Ballantine Books, 2007), vii–viii.

2. S. Begley, "Scans of Monks' Brains Show Meditation Alters Structure, Functioning," *Wall Street Journal,* November 5, 2004, http://online.wsj.com/article/0,,SB109959818932165108,00.html.

3. Mind & Life Institute, Meeting XII, 2004.

4. Norman Doidge, *The Brain That Changes Itself* (New York: Penguin, 2007).

5. G. Kempermann, H. G. Kuhn, and F. H. Gage, "More Hippocampal Neurons in Adult Mice Living in an Enriched Environment," *Nature* 386 (1997): 493–95; G. Kempermann, H. G. Kuhn, and F. H. Gage, "Experience-Induced Neurogenesis in the Senescent Dentate Gyrus," *Journal of Neuroscience* 18 (1998): 3206–12.

6. B. Oaff, "Black-Cab Drivers," *Observer,* November 10, 2002, http://www.guardian.co.uk/money/2002/nov/10/wageslaves.careers.

7. "Taxi Drivers' Brains 'Grow' on the Job," BBC News, Tuesday, March 14, 2000, http://news.bbc.co.uk/2/hi/677048.stm.

8. J. Lenzer, "Study of London Taxi Drivers Wins Ig Nobel Prize," BMJ.com, October 11, 2003, 327(7419): 831, http://www.ncbi.nlm.nih.gov/pmc.

9. V. Paquette et al., "'Change the Mind and You Change the Brain': Effects of Cognitive-Behavioral Therapy on the Neural Correlates of Spider Phobia," *Neuroimage* 18 (2003): 401–9.

10. Paquette et al., "'Change the Mind and You Change the Brain,'" 401–9.

11. M. Beauregard, J. Lévesque, and P. Bourgouin, "Neural Correlates of the Conscious Self-Regulation of Emotion," *Journal of Neuroscience* 21 (2001): RC165 (1–6).

12. J. Lévesque et al., "Neural Circuitry Underlying Voluntary Self-Regulation of Sadness," *Biological Psychiatry* 53 (2003): 502–10.

13. E. Perreau-Linck et al., "In Vivo Measurements of Brain Trapping of a-[11C]methyl-L-tryptophan During Acute Mood Changes," *Journal of Psychiatry and Neuroscience* 32 (2007): 430–34.

14. Begley, "Scans of Monks' Brains."

15. R. J. Davidson and A. Lutz, "Buddha's Brain: Neuroplasticity and Meditation," *IEEE Signal Processing Magazine* 25 (2008): 174–76.

16. A. Chiesa and A. Serretti, "A Systematic Review of Neurobiological and Clinical Features of Mindfulness Meditations," *Psychological Medicine* 40 (2010): 1239–52.

17. B. A. Wallace, "The Buddhist Tradition of Samatha: Methods for Refining and Examining Consciousness," *Journal of Consciousness Studies* 6 (1999): 175–87.

18. J. Geirland, "Buddha on the Brain—The Hot New Frontier of Neuroscience: Meditation!," *Wired,* February 2006, http://www.wired .com/wired/archive/14.02/dalai.html.

19. D. Adam, "Plan for Dalai Lama Lecture Angers Neuroscientists," *Guardian,* July 27, 2005, http://www.guardian.co.uk/world/2005/jul/27/ -research.highereducation.

20. http://www.petitiononline.com/sfn2005/.

21. Adam, "Plan for Dalai Lama Lecture."

22. D. Gilgoff, "Can Meditation Change Your Brain? Contemplative Neuroscientists Believe It Can," CNN, October 26, 2010, http://religion .blogs.cnn.com/2010/10/26/can-meditation-change-your-brain-contem plative-neuroscientists-believe-it-can/.

23. B. R. Cahn and J. Polich, "Meditation States and Traits: EEG, ERP, and Neuroimaging Studies," *Psychological Bulletin* 132 (2006): 180–211.

24. F. Travis and J. Shear, "Focused Attention, Open Monitoring and Automatic Self-Transcending: Categories to Organize Meditations from Vedic, Buddhist and Chinese Traditions," *Consciousness and Cognition* 19 (2010): 1110–18.

25. A. Lutz et al., "Long-Term Meditators Self-Induce High-Amplitude Gamma Synchrony During Mental Practice," *Proceedings of the National Academy of Sciences of the United States of America* 101 (2004): 16369–73.

26. S. Begley, "How Thinking Can Change the Brain," January 19, 2007, http://online.wsj.com/article/SB116915058061980596.html.

27. J. A. Brefczynski-Lewis et al., "Neural Correlates of Attentional Expertise in Long-Term Meditation Practitioners," *Proceedings of the National Academy of Sciences of the United States of America* 104 (2007): 11483–88.

28. A. Lutz et al., "Regulation of the Neural Circuitry of Emotion by Compassion Meditation: Effects of Meditative Expertise," *PLoS One* 3 (2008): e1897.

29. V. Taylor et al., "Impact of Mindfulness on the Neural Responses to Emotional Pictures in Experienced and Beginning Meditators," *Neuro-Image* 57 (2011): 1524–33.

30. S. W. Lazar et al., "Meditation Experience Is Associated with Increased Cortical Thickness," *Neuroreport* 16 (2005): 1893–97.

31. W. J. Cromie, "Meditation Found to Increase Brain Size," January 27, 2006, http://www.physorg.com/news10312.html.

32. B. K. Hölzel et al., "Mindfulness Practice Leads to Increases in Regional Brain Gray Matter Density," *Psychiatry Research* 191 (2011): 36–43.

33. G. Pagnoni and M. Cekic, "Age Effects on Gray Matter Volume and Attentional Performance in Zen Meditation," *Neurobiology of Aging* 28 (2007): 1623–27.

34. Y. Y. Tang et al., "Short-Term Meditation Induces White Matter Changes in the Anterior Cingulate," *Proceedings of the National Academy of Sciences of the United States of America* 107 (2010): 15649–52.

35. The Shamatha Project, http://mindbrain.ucdavis.edu/-people/jeremy/shamatha-project/.

36. O. C. Simonton, J. Creighton, and S. Matthews Simonton, *Getting Well Again: The Bestselling Classic About the Simontons' Revolutionary Life-saving Self-Awareness Techniques* (New York: Bantam, 1992.)

Chapter 4: Surfing the Psychosomatic Network

1. http://www.holisticu.org/jack/mindovermatter.html.

2. L. K. Kothari, A. Bordia, and O. P. Gupta, "The Yogi Claim of Voluntary Control over the Heart Beat: An Unusual Demonstration," *Journal of the American Heart Association* 86 (1973): 282–84.

3. L. K. Kothari et al., "A Letter to the *Journal of the American Heart Association*," 1973, http://www.sol.com.au/kor/10_02.htm.

4. Collagen is the fibrous tissue that binds together the cells of the body.

5. Norman Cousins, *Anatomy of an Illness as Perceived by the Patient: Reflections on Healing and Regeneration* (New York: Norton, 1979), 33.

6. Cousins, *Anatomy of an Illness*.

7. Tom Long, "Jordan Fieldman, 38; Doctor Promoted Holistic Medicine," *Boston Globe,* June 10, 2004, http://www.boston.com/news/globe/obituaries/articles/2004/06/10/jordan_fieldman_38_doctor_promoted_holistic_medicine/.

8. http://www.berniesiegelmd.com/remarkable_recoveries.htm.

9. As noted in the introduction, the Greek physicians of antiquity saw the mind and body as one. They thought that mind plays an important role in health and illness. Along the same lines, the medieval Islamic physicians Ahmed ibn Sahl al-Balkhi and Haly Abbas proposed that illness sometimes reflects the effect of mind on the body. D. Nurdeen and A. T.

Mansor, "Mental Health in Islamic Medical Tradition," *International Medical Journal* 4 (2005): 76–79.

Western scientific and technological progress in the nineteenth century allowed scientists to peer into blood and water and discover the teeming life that was invisible to the unaided eye. As they began to identify various agents of diseases, such as bacteria, fungus, and viruses, physicians understandably began to believe that the treatment of infectious diseases required only the elimination of such agents. This belief, in tandem with the progressive character of the Industrial Revolution, led them to adopt the mechanistic view. They saw the body as a biological machine that cannot be influenced by mind and conceived of medicine's role as essentially to repair bodily malfunctions. E. M. Sternberg and P. W. Gold, *The Mind-Body Interaction in Disease,* http://being.publicradio.org/programs/stress/mindbodyessay.shtml.

10. E. M. Sternberg, *The Balance Within: The Science Connecting Health and Emotions* (New York: W. H. Freeman, 2001).

11. G. F. Solomon and R. H. Moss, "Emotions, Immunity, and Disease: A Speculative Theoretical Integration," *Archives of General Psychiatry* 11 (1964): 657–74.

12. J. S. Gordon, "Mind-Body Medicine and Cancer," *Hematology/Oncology Clinics of North America* 22 (2008): 683–708.

13. J. M. Williams et al., "Sympathetic Innervation of Murine Thymus and Spleen: Evidence for a Functional Link Between the Nervous and Immune Systems," *Brain Research Bulletin* 6 (1981): 83–94.

14. Candace Pert, *Molecules of Emotion: The Science Behind Mind-Body Medicine* (New York: Touchstone, 1997).

15. O. Ray, "The Revolutionary Health Science of Psychoendoneuroimmunology," *Annals of the New York Academy of Sciences* 1032 (2004): 35–51.

16. Pert, *Molecules of Emotion.*

17. Ray, "Psychoendoneuroimmunology."

18. J. A. Dusek et al., "Genomic Counter-Stress Changes Induced by the Relaxation Response," *PLoS One* 3 (2008): e2576.

19. Ray, "Psychoendoneuroimmunology."

20. Sternberg and Gold, *Mind-Body Interaction in Disease.*

21. Ray, "Psychoendoneuroimmunology."

22. L. Vitetta et al., "Mind–Body Medicine: Stress and Its Impact on Overall Health and Longevity," *Annals of the New York Academy of Sciences* 1057 (2005): 492–505.

23. A. Baum, L. Cohen, and M. Hall, "Control and Intrusive Memories as Possible Determinants of Chronic Stress," *Psychosomatic Medicine* 55 (1993): 274–86.

24. S. A. Everson et al., "Hopelessness and Risk of Mortality and Incidence of Myocardial Infarction and Cancer," *Psychosomatic Medicine* 58 (1996): 113–21.

25. E. W. Kelly, "Psychophysiological Influence," in *Irreducible Mind: Toward a Psychology for the Twenty-First Century,* ed. E. F. Kelly and E. W. Kelly (Lanham, MD: Rowman & Littlefield, 2007), 117–239.

26. M. Callanan and P. Kelley, *Final Gifts: Understanding the Special Needs, Awareness, and Communications of the Dying* (New York: Bantam, 1993).

27. Kelly, "Psychophysiological Influence."

28. S. Cohen et al., "Emotional Style and Susceptibility to the Common Cold," *Psychosomatic Medicine* 65 (2003): 652–57; S. Cohen et al., "Positive Emotional Style Predicts Resistance to Illness After Experimental Exposure to Rhinovirus or Influenza A Virus," *Psychosomatic Medicine* 68 (2006): 809–15.

29. S. Greer, "Psychological Response to Cancer and Survival," *Psychological Medicine* 21 (1991): 43–49.

30. NK cells are crucially involved in the rejection of tumors and cells infected by viruses.

31. J. K. Kiecolt-Glaser et al., "Psychoneuroimmunology and Psychosomatic Medicine: Back to the Future," *Psychosomatic Medicine* 64 (2002): 15–28.

32. Gordon, "Mind-Body Medicine and Cancer."

33. Gordon, "Mind-Body Medicine and Cancer."

34. E. C. Trakhtenberg, "The Effects of Guided Imagery on the Immune System: A Critical Review," *International Journal of Neuroscience* 118 (2008): 839–55.

35. O. C. Simonton, J. Creighton, and S. Matthews Simonton, *Getting Well Again: The Bestselling Classic About the Simontons' Revolutionary Lifesaving Self-Awareness Techniques* (New York: Bantam, 1992).

36. Gordon, "Mind-Body Medicine and Cancer."

37. L. E. Carlson and B. D. Bultz, "Mind-Body Interventions in Oncology," *Current Treatment Options in Oncology* 9 (2008): 127–34.

38. Kelly, "Psychophysiological Influence."

39. M. Falco, "Herschel Walker Reveals Many Sides of Himself," CNN, April 15, 2008, http://articles.cnn.com/2008–04–15/health/herschel .walker.did_1_disorder-russian-roulette-herschel-walker/3?_s=P.M.: HEALTH.

40. Falco, Herschel Walker; H. Walker and J. Mungadze, *Breaking Free: My Life with Dissociative Identity Disorder* (New York: Touchstone, 2008).

41. Falco, "Herschel Walker."

42. Falco, "Herschel Walker."

43. A. A. Reinders et al., "One Brain, Two Selves," *Neuroimage* 20 (2003): 2119–25.

44. Kelly, "Psychophysiological Influence," 117–239.

45. H. Benson et al., "Body Temperature Changes During the Practice of g Tummo Yoga," *Nature* 295 (1982): 234–36.

46. Helen Flanders Dunbar, *Emotions and Bodily Changes,* 4th ed. (New York: Columbia Univ. Press, 1954).

47. Elmer Green and Alyce Green, *Beyond Biofeedback* (New York: Delacorte Press/S. Lawrence, 1977).

48. Ray, "Psychoendoneuroimmunology," 1.

49. Kelly, "Psychophysiological Influence."

50. Pert, *Molecules of Emotion.*

51. Kelly, "Psychophysiological Influence."

Chapter 5: The Mind Force Within

1. L. Alderman, "Using Hypnosis to Gain More Control Over Your Illness," *New York Times,* April 16, 2011, http://www.nytimes .com/2011/04/16/health/16patient.html.

2. A. A. Mason, "A Case of Congenital Ichthyosiform Erythrodermia of Brocq Treated by Hypnosis," *British Medical Journal* 2 (1952): 422.

3. Mason, "Congenital Ichthyosiform Erythrodermia," 422–23.

4. A. A. Mason, "Icthyosis and Hypnosis," 1955, http://www.ncbi.nlm .nih.gov/pmc/articles/P.M.C1980190/?page=2.

5. D. Barrett, "The Power of Hypnosis," *Psychology Today,* January/ February, 2001, http://www.psychologytoday.com/articles/200101/the -power-hypnosis.

6. J. H. Stewart, "Hypnosis in Contemporary Medicine," *Mayo Clinic Proceedings* 80 (2005): 511–24.

7. J. Esdaile, *Mesmerism in India, and Its Practical Applications in Surgery and Medicine* (London: Longman, Brown, Green, and Longman, 1846), 240–41.

8. Stewart, "Hypnosis in Contemporary Medicine."

9. Alderman, "Using Hypnosis."

10. D. Elkan, "You Will Feel Nothing: Hypnosis Meets Surgery," *HypnoGenesis,* 2006, http://www.hypnos.co.uk/hypnomag/hypnosisnews/ -hypnosissurgery.htm.

11. M. Cheng, "For Some, Hypnosis Eases Pain, Recovery of Surgery," Associated Press, July 27, 2011, http://www.cnsnews.com/news/article/ -some-hypnosis-eases-pain-recovery-surgery.

12. Cheng, "For Some, Hypnosis Eases Pain."

13. G. H. Montgomery, K. N. DuHamel, and W. H. Redd, "A Meta-analysis of Hypnotically Induced Analgesia: How Effective Is Hypnosis?," *International Journal of Clinical and Experimental Hypnosis* 48 (2000): 138–53.

14. M. M. Berger, et al., "Impact of a Pain Protocol Including Hypnosis in Major Burns," *Burns* 36 (2010): 639–46.

15. G. R. Elkins et al., "Hypnosis to Reduce Pain in Cancer Survivors with Advanced Disease: A Prospective Study," *Journal of Cancer and Integrative Medicine* 2 (2004): 167–72.

16. J. A. D. Anderson, M. A. Basker, and R. Dalton, "Migraine and Hypnotherapy," *International Journal of Clinical and Experimental Hypnosis* 23 (1975): 48–58.

17. A. S. Landolt and L. S. Milling, "The Efficacy of Hypnosis as an Intervention for Labor and Delivery Pain: A Comprehensive Methodological Review," *Clinical Psychology Review* 3 (2011): 1022–31.

18. T. Harmon, M. Hynan, and T. Tyre, "Improved Obstetric Outcomes Using Hypnotic Analgesia and Skill Mastery Combined with Childbirth Education," *Journal of Consulting and Clinical Psychology* 58 (1990): 525–30.

19. L. Mehl-Madrona, "Hypnosis to Facilitate Uncomplicated Birth," *American Journal of Clinical Hypnosis* 46 (2004): 299–312.

20. A. M. Cyna, M. I. Andrew, and G. L. McAuliffe, "Antenatal Self-Hypnosis for Labor and Childbirth: A Pilot Study," *Anesthesia Intensive Care* 34 (2006): 464–69.

21. H. Saadat et al., "Hypnosis Reduces Preoperative Anxiety in Adult Patients," *Anesthesia and Analgesia* 102 (2006): 1394–96.

22. E. V. Lang et al., "Adjunctive Self-Hypnotic Relaxation for Outpatient Medical Procedures: A Prospective Randomized Trial with Women Undergoing Large Core Breast Biopsy," *Pain* 126 (2006): 155–64.

23. P. D. Shenefelt, "Hypnosis in Dermatology," *Archives of Dermatology* 136 (2000): 393–99.

24. Shenefelt, "Hypnosis in Dermatology."

25. C. Ginandes et al., "Can Medical Hypnosis Accelerate Post-Surgical Wound Healing? Results of a Clinical Trial," *American Journal of Clinical Hypnosis* 45 (2003): 333–51.

26. D. M. Ewin, "Hypnosis in Burn Therapy," in *Hypnosis,* ed. G. Burrows, D. R. Collison, and L. Dennerstein (Amsterdam: Elsevier, 1979), 269–75; Shenefelt, "Hypnosis in Dermatology."

27. Stewart, "Hypnosis in Contemporary Medicine."

28. R. D. Willard, "Breast Enlargement Through Visual Imagery and Hypnosis," *American Journal of Clinical Hypnosis* 19 (1977): 195–200.

29. C. S. Ginandes and D. I. Rosenthal, "Using Hypnosis to Accelerate

the Healing of Bone Fractures: A Randomized Controlled Pilot Study," *Alternative Therapies, Health and Medicine* 5 (1999): 67–75.

30. D. Barrett, "Trance-Related Pseudocyesis in a Male," *International Journal of Clinical and Experimental Hypnosis* 36 (1988): 256–61.

31. S. M. Kosslyn et al., "Hypnotic Visual Illusion Alters Color Processing in the Brain," *American Journal of Psychiatry* 157 (2000): 1279–84.

32. Kosslyn et al., "Hypnotic Visual Illusion."

33. S. Schulz-Stübner et al., "Clinical Hypnosis Modulates Functional Magnetic Resonance Imaging Signal Intensities and Pain Perception in a Thermal Stimulation Paradigm," *Regional Anesthesia and Pain Medicine* 29 (2004): 549–56.

34. M. Pyka et al., "Brain Correlates of Hypnotic Paralysis: A Resting-State fMRI Study," *Neuroimage* 56 (2011): 2173–82.

35. Pyka et al., "Brain Correlates of Hypnotic Paralysis."

36. E. W. Kelly, "Psychophysiological Influence," in *Irreducible Mind: Toward a Psychology for the Twenty-First Century,* ed. E. F. Kelly and E. W. Kelly (Lanham, MD: Rowman & Littlefield, 2007), 117–239.

37. Kelly, "Psychophysiological Influence."

Chapter 6: Beyond Space and Time

1. http://www.goodreads.com/author/quotes/6172.S_ren_Kierkegaard.

2. Joe McMoneagle, *Mind Trek: Exploring Conciousness, Time, and Space Through Remote Viewing* (Norfolk, VA: Hampton Roads, 1993), 29–30.

3. McMoneagle, *Mind Trek,* 29–30.

4. Dean Radin, *The Conscious Universe: The Scientific Truth of Psychic Phenomena* (New York: HarperEdge, 1997).

5. Radin, *Conscious Universe.*

6. J. Utts, "An Assessment of the Evidence for Psychic Functioning," *Journal of Scientific Exploration* 10 (1996): 3.

7. Radin, *Conscious Universe.*

8. R. Milton, *Alternative Science* (Rochester, VT: Park Street Press, 1996).

9. Chris Carter, *Parapsychology and the Skeptics: A Scientific Argument for the Existence of ESP* (Pittsburgh, PA: Sterling House, 2007).

10. Carter, *Parapsychology and the Skeptics.*

11. Radin, *Conscious Universe.*

12. C. Honorton and E. I. Schecter, "Ganzfeld Target Retrieval with an Automated System: A Model for Initial Ganzfeld Success," in *RIP 1986,* ed. D. B. Weiner and R. D Nelson (Metuchen, NJ: Scarecrow Press, 1987), 36–39.

13. Dean Radin, *Entangled Minds: Extrasensory Experiences in a Quantum Reality* (New York: Pocket Books, 2006).

14. Honorton and Schecter, "Ganzfeld Target Retrieval."

15. Radin, *Conscious Universe.*

16. R. Rosenthal, "The File Drawer Problem and Tolerance for Null Results," *Psychological Bulletin* 86 (1979): 638–41.

17. Radin, *Entangled Minds.*

18. Radin, *Conscious Universe.*

19. P. Tressoldi, L. Storm, and D. Radin, "Extrasensory Perception and Quantum Models of Cognition," *NeuroQuantology* 8 (2010): S81–S87.

20. N. J. Holt, "Are Artistic Populations Psi-Conducive? Testing the Relationship Between Creativity and Psi with an Experience-Sampling Protocol," in *Proceedings of the Fiftieth Annual Convention of the Parapsychological Association* (Petaluma, CA: Parapsychological Association, 2007), 31–47.

21. C. Honorton and D. C. Ferrari, "Future Telling: A Meta-analysis of Forced-Choice Precognition Experiments, 1935–1987," *Journal of Parapsychology* 53 (1989): 281–308.

22. Radin, *Conscious Universe.*

23. Radin, *Conscious Universe.*

24. D. J. Bierman and H. S. Scholte, "Anomalous Anticipatory Brain Activation Preceding Exposure of Emotional and Neutral Pictures," paper presented at Toward a Science of Consciousness IV, Tucson, AZ, 2002.

25. Dean Radin, personal communication.

26. D. Bem, "Feeling the Future: Experimental Evidence for Anomalous Retroactive Influences on Cognition and Affect," *Journal of Personality and Social Psychology* 100 (2011): 407–25.

27. Bem, "Feeling the Future," 409.

28. Bem, "Feeling the Future."

29. E. Halliwell, "Can We Feel the Future Through Psi? Don't Rule It Out," *Guardian* (2011), http://www.guardian.co.uk/commentisfree/-belief/2011/jan/25/precognition-feeling-the-future.

30. Jim Schnabel, "Don't Mess with My Reality," January 16, 2011, http://hereticalnotions.com/2011/01/16/dont-mess-with-my-reality/.

31. Radin, *Conscious Universe.*

32. Radin, *Conscious Universe.*

33. R. D. Nelson and D. Radin, "When Immovable Objections meet Irresistible Evidence," *Behavioral and Brain Sciences* 10 (1987): 600–601.

34. R. Jahn et al., "Mind/Machine Interaction Consortium: PortREG Replication Experiments," *Journal of Scientific Exploration* 14 (2000): 499–555.

35. Radin, *Conscious Universe.*

36. Radin, *Conscious Universe.*

37. W. G. Braud and M. J. Schlitz, "Consciousness Interactions with Remote Biological Systems: Anomalous Intentionality Effects," *Subtle Energies* 2 (1991): 1–46.

38. T. D. Duane and R. Behrendt, "Extrasensory Electroencephalographic Induction Between Identical Twins," *Science* 150 (1965): 367.

39. L. J. Standish et al., "Electroencephalographic Evidence of Correlated Event-Related Signals Between the Brains of Spatially and Sensory Isolated Human Subjects," *Journal of Alternative and Complementary Medicine* 10 (2004): 307–14.

40. T. L. Richards et al., "Replicable functional magnetic resonance imaging evidence of correlated brain signals between physically and sensory isolated subjects," *Journal of Alternative and Complementary Medicine* 11 (2005): 955–63.

41. Radin, *Conscious Universe;* Radin, *Entangled Minds.*

42. Radin, *Conscious Universe;* Radin, *Entangled Minds.*

Chapter 7: Mind Out of Body

1. R. Haynes, *The Society for Psychical Research, 1882–1982, A History* (London: Macdonald & Co., 1982), 83.

2. The brain stem plays an important role in the regulation of cardiac and respiratory functions. This cerebral structure is also critical in maintaining consciousness.

3. Michael B. Sabom, *Light and Death: One Doctor's Fascinating Account of Near-Death Experiences* (Grand Rapids, MI: Zondervan, 1998).

4. Sabom, *Light and Death,* 41.

5. Sabom, *Light and Death,* 42.

6. Sabom, *Light and Death,* 42.

7. *The Day I Died: The Mind, the Brain, and Near-Death Experiences.* BBC, 2002.

8. *The Day I Died: The Mind, the Brain, and Near-Death Experiences.* BBC, 2002.

9. Raymond Moody Jr., *Life After Life* (Covington, GA: Mockingbird Books, 1975).

10. Clinical death is determined by either the cessation of heart beat and breathing or the irreversible end of all brain activity.

11. Pim van Lommel, *Consciousness Beyond Life: The Science of the Near-Death Experience* (New York: HarperCollins, 2010).

12. Pim van Lommel, *Consciousness Beyond Life.*

13. Pim van Lommel, *Consciousness Beyond Life.*

14. Kenneth Ring and Evelyn Elsaesser-Valarino, *Lessons from the Light: What We Can Learn from the Near-Death Experience* (New York and London: Plenum, Insight Books, 1998).

15. A. Kellehear, "Culture, Biology and the Near-Death Experience," *Journal of Nervous and Mental Disease* 181 (1993): 148–56.

16. Ring and Elsaesser-Valarino, *Lessons from the Light.*

17. Reinée-Pasarow, "A Personal Account of an NDE," *Vital Signs* 1, no. 3 (1981): 4–7.

18. http://www.iands.org/nde_archives/experiencer_accounts/chose _second_chance.html.

19. van Lommel, *Consciousness Beyond Life.*

20. P. van Lommel et al., "Near-Death Experience in Survivors of Cardiac Arrest: A Prospective Study in the Netherlands," *Lancet* 358 (2001): 2039–45.

21. van Lommel. *Consciousness Beyond Life;* Ring and Elsaesser-Valarino, *Lessons from the Light.*

22. van Lommel, *Consciousness Beyond Life.*

23. B. Greyson and N. E. Bush, "Distressing Near-Death Experiences," *Psychiatry* 55 (1992): 95–110.

24. E. W. Kelly, B. Greyson, and E. D. Kelly. "Unusual Experiences Near Death and Related Phenomena," in *Irreducible Mind: Toward a Psychology for the Twenty-First Century,* ed. E. F. Kelly and E. W. Kelly (Lanham, MD: Rowman & Littlefield, 2007), 367–422.

25. K. Clark, "Clinical Interventions with Near-Death Experiencers," in *The Near-Death Experience: Problems, Prospects, Perspectives,* ed. B. Greyson and C. P. Flynn (Springfield, IL: Charles C. Thomas, 1984).

26. Clark, "Clinical Interventions," 243.

27. H. L. Clute and W. J. Levy, "Electroencephalographic Changes During Brief Cardiac Arrest in Humans," *Anesthesiology* 73 (1990): 821–25.

28. S. Parnia et al., "A Qualitative and Quantitative Study of the Incidence, Features and Aetiology of Near Death Experiences in Cardiac Arrest Survivors," *Resuscitation* 48 (2001): 149–56.

29. P. van Lommel et al., "Near-Death Experience in Survivors of Cardiac Arrest."

30. Parnia et al., "Incidence, Features and Aetiology of Near Death Experiences."

31. B. Greyson, "Incidence and Correlates of Near-Death Experiences in a Cardiac Care Unit," *General Hospital Psychiatry* 25 (2003): 269–76.

32. E. W. Cook, B. Greyson, and I. Stevenson, "Do Any Near-Death Experiences Provide Evidence for the Survival of Human Personality After

Death? Relevant Features and Illustrative Case Reports," *Journal of Scientific Exploration* 12 (1998): 377–406.

33. van Lommel et al., "Near-Death Experience in Survivors of Cardiac Arrest."

34. K. Ring and S. Cooper, "Near-Death and Out-of-Body Experiences in the Blind: A Study of Apparent Eyeless Vision," *Journal of Near-Death Studies* 16 (1997): 101–47.

35. Ring and Elsaesser-Valarino, *Lessons from the Light*, 87–90.

36. Ring and Cooper, "Near-Death and Out-of-Body Experiences in the Blind."

37. Ring and Cooper, "Near-Death and Out-of-Body Experiences in the Blind."

38. Ring and Cooper, "Near-Death and Out-of-Body Experiences in the Blind."

39. O. Blanke et al., "Stimulating Illusory Own-Body Perceptions," *Nature* 419 (2002): 269–70.

40. F. Grace, *Out of Body Experiences Explained*, CBS London, September 19, 2002.

41. O. Blanke et al., "Out-of-Body Experience and Autoscopy of Neurological Origin," *Brain* 127(Pt 2) (2004): 243–58.

42. van Lommel, *Consciousness Beyond Life*.

43. Grace. *Out of Body Experiences*.

44. S. J. Blackmore and T. Troscianko, "The Physiology of the Tunnel," *Journal of Near-Death Studies* 8 (1988): 15–28.

45. van Lommel et al., "Near-Death Experience in Survivors of Cardiac Arrest."

46. Sam Parnia, *What Happens When We Die: A Ground-Breaking Study into the Nature of Life and Death* (London: Hay House, 2008).

47. J. E. Whinnery, "Psychophysiologic Correlates of Unconsciousness and Near-Death Experiences," *Journal of Near-Death Studies* 5 (1997): 232–58.

48. K. Jansen, "Near Death Experience and the NMDA Receptor," *British Medical Journal* 298 (1989): 1708.

49. M. A. Persinger, "Near-Death Experiences and Ecstasy: A Product of the Organization of the Human Brain?," in *Mind Myths: Exploring Popular Assumptions About the Mind and Brain,* ed. S. Della Sala (Chichester, UK: Wiley, 1999), 85–99.

50. Kelly et al., "Unusual Experiences."

51. E. J. Larson and L. Witham, "Leading Scientists Still Reject God," *Nature* 394 (1998): 313.

Chapter 8: Embracing a Greater Self

1. William Blake, *The Marriage of Heaven and Hell* (New York: Oxford Univ. Press, 1975).

2. Richard M. Bucke, *Cosmic Consciousness* (New York: Dutton, 1969). (Original work published 1901)

3. S. Grof and H. Z. Bennett, *The Holotropic Mind: The Three Levels of Human Consciousness and How They Shape Our Lives* (New York: Harper Collins, 1992), 97–98.

4. http://www.issc-taste.org/arc/dbo.cgi?set=expom&id=00004&ss=1.

5. Grof and Bennett, *The Holotropic Mind.*

6. Walter T. Stace, *Mysticism and Philosophy* (New York: Macmillan, 1960).

7. Bucke, *Cosmic Consciousness.*

8. http://www.issc-taste.org/arc/dbo.cgi?set=expom&id=00004&ss=1.

9. Evelyn Underhill, *Mysticism: A Study in the Nature and Development of Man's Spiritual Consciousness* (New York: New American Library, 1974).

10. J. L. Waldron, "The Life Impact of Transcendent Experiences with a Pronounced Quality of Noesis," *Journal of Transpersonal Psychology* 30 (1998): 103–34; J. Levin and L. Steele, "The Transcendent Experience: Conceptual, Theoretical, and Epidemiological Perspectives," *Explore* 1 (2005): 89–101.

11. Grof and Bennett, *Holotropic Mind.*

12. Andrew M. Greeley, *The Sociology of the Paranormal: A Reconnaissance* (Beverly Hills, CA: Sage, 1975).

13. L. Thomas and P. Cooper, "Incidence and Psychological Correlates of Intense Spiritual Experiences," *Journal of Transpersonal Psychology* 12 (1980): 75–85.

14. E. F. Kelly and M. Grosso, "Mystical Experience," in *Irreducible Mind: Toward a Psychology for the Twenty-First Century,* ed. E. F. Kelly and E. W. Kelly (Lanham, MD: Rowman & Littlefield, 2007).

15. J. L. Saver and J. Rabin, "The Neural Substrates of Religious Experience," *Journal of Neuropsychiatry and Clinical Neuroscience* 9 (1997): 498–510.

16. T. Alajouanine, "Dostoiewski's Epilepsy," *Brain* 86 (1963): 209–18.

17. J. R. Hughes, "Did All Those Famous People Really Have Epilepsy?," *Epilepsy and Behavior* 6 (2005): 115–39.

18. D. Hay, "'The Biology of God': What Is the Current Status of Hardy's Hypothesis?," *International Journal for the Psychology of Religion* 4 (1994): 1–23.

19. Mario Beauregard and Denyse O'Leary, *The Spiritual Brain* (New York: HarperCollins, 2007).

20. M. A. Persinger, "Religious and Mystical Experiences as Artefacts

of Temporal Lobe Function: A General Hypothesis," *Perceptual and Motor Skills* 57 (1983): 1255–62.

21. M. A. Persinger and F. Healey, "Experimental Facilitation of the Sensed Presence: Possible Intercalation Between the Hemispheres Induced by Complex Magnetic Fields," *Journal of Nervous and Mental Diseases* 190 (2002): 533–41.

22. P. Granqvist et al., "Sensed Presence and Mystical Experiences Are Predicted by Suggestibility, Not by the Application of Transcranial Weak Complex Magnetic Fields," *Neuroscience Letters* 379 (2005): 1–6.

23. E. Halgren et al., "Mental Phenomena Evoked by Electrical Stimulation of the Human Hippocampal Formation and Amygdala," *Brain* 101 (1978): 83–117.

24. A. Newberg et al., "Cerebral Blood Flow During Meditative Prayer: Preliminary Findings and Methodological Issues," *Perceptual and Motor Skills* 97 (2003): 625–30.

25. M. Beauregard and V. Paquette, "Neural Correlates of a Mystical Experience in Carmelite Nuns," *Neuroscience Letters* 405 (2006): 186–90.

26. M. Beauregard, C. Courtemanche, and V. Paquette, "Brain Activity in Near-Death Experiencers During a Meditative State," *Resuscitation* 80 (2009): 1006–10.

27. Kelly and Grosso, "Mystical Experience."

28. Gordon Wasson, *Soma: Divine Mushroom of Immortality* (New York: Harcourt Brace & World, 1968).

29. Aldous Huxley, *The Doors of Perception* (New York: Harper & Row, 1954).

30. Huston Smith, *The Religions of Man* (New York: Harper, 1958).

31. Huston Smith, *Empirical Metaphysics,* http://www.psychedelic-library.org/books/ecstatic5.htm.

32. As an individual who has undergone so-called spontaneous MEs (including states of Cosmic Consciousness) and spiritual experiences occasioned by the ingestion of psychedelic substances, I can attest that it can be sometimes difficult to distinguish, from an experiential perspective, these two categories of experiences.

33. Stace, *Mysticism and Philosophy.*

34. R. R. Griffiths, W. A. Richards, U. McCann, R. Jesse, "Psilocybin Can Occasion Mystical-Type Experiences Having Substantial and Sustained Personal Meaning and Spiritual Significance," *Psychopharmacology* 187 (2006): 268–83

35. R. R. Griffiths, W. A. Richards, M. Johnson, U. McCann, R. Jesse, "Mystical–Type Experiences Occasioned by Psilocybin Mediate the Attribution of Personal Meaning and Spiritual Significance 14 Months Later," *Journal of Psychopharmacology* 22 (2008): 621–32.

36. Tierney J. "Hallucinogens Have Doctors Tuning In Again," *New York Times,* April 11, 2010.

37. Grof and Bennett, *Holotropic Mind.*

Conclusion: A Great Shift in Consciousness

1. James Hopwood Jeans, *The Mysterious Universe* (Cambridge, UK: Cambridge Univ. Press, 1930), 137.

2. B. Greyson, "Cosmological Implications of Near-Death Experiences," *Journal of Cosmology* 14 (2011), http://journalofcosmology.com/Consciousness129.html.

3. Neuroscience methods allow researchers to measure physical and chemical correlates of mental events, not the mental events themselves. *Correlations between mental activity and brain activity do not imply causation and identity.* For instance, neuroscientists can record using EEG brain electrical activity from the scalp of human volunteers while they are experiencing an emotion. But the EEG activity recorded is totally different from the feelings experienced by the volunteers. And the correlation between this electrical activity and the change in the emotional state of the volunteers does not mean that the alterations in the EEG are the cause of the feelings.

The mistaken belief that mental events are identical with their neural correlates is still entertained by some science writers, journalists, and neuroscientists, and it leads to what has been called the "mereological fallacy": the erroneous attribution of mental properties to parts of the brain (or to the brain itself—see M. R. Bennett and P. M. S. Hacker, *Philosophical Foundations of Neuroscience,* New York, Blackwell, 2003). As research psychologist and philosopher of science Saulo de Freitas Araujo remarks, the materialists committing this mistake elevate the brain to an omnipotent physical entity, and seek in its properties the basic explanation of all mental phenomena. They do not appreciate that whole persons—and not only regions of the brain—are conscious, perceive, think, feel, believe, and make decisions.

4. http://www.word-gems.com/mind.html.

5. Fritjof Capra, *The Tao of Physics* (Boston: Shambhala, 1975).

6. http://quantumenigma.com/nutshell/notable-quotes-on-quantum physics/?phpMyAdmin=54029d98ba071eec0c69ff5c106b9539]

7. B. Rosenblum and F. Kuttner, *Quantum Enigma: Physics Encounters Consciousness* (New York: Oxford Univ. Press, 2006).

8. Wolfgang Pauli, *The Influence of Archetypal Ideas on the Scientific Theories of Kepler: The Interpretation of Nature and the Psyche* (London: Routledge & Kegan Paul, 1955).

9. M. Kafatos and R. Nadeau, *The Conscious Universe: Parts and Wholes in Physical Reality* (New York: Springer, 1999).

10. But Saulo de Freitas Araujo has shown that this prophetic belief was already professed by proponents of materialism in the eighteenth century. S. De Freitas Araujo, "Materialism's Eternal Return: Recurrent Patterns of Materialistic Explanations of Mental Phenomena," in *Exploring the Frontiers of the Mind-Brain Relationship,* ed. A. Moreira-Almeida and F. Santana Santos (New York: Springer, in press).

11. The discoveries of QM and the findings examined in this book let us catch a glimpse of this emerging model, about which a few authors have already written. Interested readers may consult the writings of Willis Harman, Richard Tarnas, Peter Russell, Fritjof Capra, and David Lorimer. As science is in a continuous state of flux and development, this model is by definition incomplete and temporary—and, of course, we must always remember that model of reality is not reality itself.

INDEX